PATTERNS

Ivan Gilbert, M.D.

PATTERNS

Published by Paradigm Press

Printed in the United States of America

ISBN: 978-1-312-18994-2

1 2 3

Introduction

PATTERNS IS IVAN GILBERT'S baby, the result of an unusually long period of gestation—around ninety years, I'd say. By the time his pregnancy reached full term, Ivan Gilbert had endured stern injunctions to lose weight, many disappointing false labors, even painful surgery. Yet all that was forgotten when he gazed on the outcome, and like any proud mother, fell in love.

I first met Ivan in 1993, while I was still living in Columbus, after somebody at Ohio State advised him that I might be able to edit his manuscript. He arrived at my house bearing the manuscript and two enormous cardboard boxes marked PATTERNS. A quick scan of their contents revealed hundreds and hundreds of pages of finished and unfinished articles and essays, philosophical ramblings, random thoughts, scientific disquisition, a couple of old electric bills, some ancient potato chip remains, and the musty smell of typewriter ink and paper stored too long in a dark place. There was no discernible principle of organization. My heart sank.

It rose again, however, when I took up the manuscript. Ivan said it was the work of a previous editor. (In fact there had been numerous previous editors.) What a relief it was to find a table of contents, separate chapters, and consecutive page numbers. Perhaps someone had brought order out of chaos.

"But I don't like it," Ivan said darkly.

"Why not?"

"I don't know, but I don't like it. It doesn't sound like me. Just read it and tell me what you think."

He left, and I settled in to read the manuscript. It was tough going. The only patterns I could make out were circles—circles within circles, circles intersecting circles, circles winding in and out of themselves. The tangles were understandable—those two huge boxes were squatting on my living room rug like reproachful toads demanding to be transformed into princes—but unreadable.

Further investigation into the boxes convinced me that I wouldn't be able to handle the job any better than this most recent editor, so Ivan shopped it to someone else. We became friends, though; it was a meeting of like minds.

A year later I moved to New Mexico, and even though we never met again after that year, the friendship continued by telephone until Ivan died, at ninety-one. Working from New Mexico, I edited another of his books, *My Brother Married Groucho Marx*. Then, four or five years ago, Ivan called to bemoan the problems he was encountering with Patterns.

Yet another editor had taken it on and had produced a finished manuscript, but Ivan didn't like it. It didn't sound like him. Would I just read it and tell him what I thought? By now I was quite familiar with what Ivan "sounded like," so, sure I would.

Shortly the FedEx man arrived at the door, handed me the manuscript in a fat manila envelope, then staggered inside with two enormous boxes and dropped them on the living room rug.

No. No. It couldn't be. But it was. The toads.

I read the manuscript with pleasure; the writing was clear and orderly, and it seemed to me the editor had done a fine, professional job. Ivan disagreed. It was too dry, he said, too much like a science textbook, and that was not at all what he wanted his book to be. He wanted it to sound like him, to be interesting. And time had become a factor: he wanted it finished before he died—an event that, now in his late eighties, he had begun grudgingly to acknowledge might come to pass someday. Could I please try it? Well, gosh, Ivan, uh, I guess so, well, all right.

My initial editorial ploy was to haul and push and shove and ram the toads into a closet and shut the door. The second was to re-read what I still considered a very good manuscript, and then to stare at it for a long time. Maybe I could use its chapter headings as an outline and add some examples to illustrate its major points… but how on earth to make it sound like Ivan?

No idea. More staring ensued.

This was getting nowhere, so I chose a chapter at random, took a deep breath, dived in without knowing how deep the water was, and *lo!*, somehow the problem solved itself. It was as though Ivan were in the room with me, as though I were channeling him. Twenty years' worth of telephone conversations was paying off. Or so I hoped—but only faintly, because Ivan the Unsatisfied had yet to weigh in. I e-mailed the edited chapter to him and waited for developments.

Next day he called. By now I was certain he'd hate it.

"I love it," he said.

At long last, *Patterns* had found an editorial home.

Over the time that I worked with Ivan to complete the book, I came to understand it as a final distillation of what he had experienced and learned and wondered about and hoped for and concluded throughout his long life.

What *Patterns* positively isn't—he insisted on this—is pedantic. He loved to explain difficult concepts in ways anybody could understand, and he loathed stuffiness. Like

Ivan himself, *Patterns* is smart, funny, controversial, obstinate, exasperating, whimsical, eclectic.

The reason he finally entrusted his book to me was that, he said, I had captured his voice.

I'm pleased to believe that is true.

Gail Burke
2014

Gail Burke is a professor of English at New Mexico State University, Grants. She taught English at The Ohio State University prior to moving to New Mexico and has edited several books, including Ivan Gilbert's My Brother Married Groucho Marx.

To Begin with. . .

AS I HAVE done so often for thirty-five years, I'm relaxing in my garden. Everything is in bloom. The crisp young summer of June has mutated into lush and heavily fecund August. With the grass in full growth, the bushes overgrown, the flowers peaking or past their bloom, the atmosphere is overripe. Pollen has done its work most of the summer; now the garden is indolent, bending with ripeness—a woman mature and beautiful, content and at peace with herself.

I'm thinking about this garden because I have been thinking about patterns. The garden was this way last August and August twelve years ago, and it will be this way twenty years from now.

The garden is a pattern full of patterns.

Each tree, bush, flower, and blade of grass has its own pattern that begins with a seed and ends with the plant's death, and each pattern alters in accord with the plant's heredity and its environment, which dictate its color, its height, its time to flower and its time to die. The sum of all these patterns is the garden's pattern, repeating from year to year, cycling and recycling.

This is the garden's life.

Now, at a mature age, I understand that patterns form and summarize all of life. Patterns have shaped what I, and you, have done and become. How do they develop? What do they control? How do they fade and change? What do they explain?

Five years ago, sitting here in the garden, I began this book. The garden has not changed much in that time. So that we could build a new kitchen addition, our red-brick circular patio was replaced by a red-brick square patio. We had to take down one of the tall old oaks, and that left a ragged tear in my tree canopy, but very soon new growth from the other trees began to fill it in. The hawthorn tree by the window is gone too; we removed it to extend the kitchen. Fewer cardinals visit us now, for they love hawthorn berries. But the garden remains, surviving winter snows, icy colds, droughts, infestations of vermin. We have seen some loss here and some gain there, but the beauty and the structure and the progression trace their patterns through the seasons.

Our trees are tall and short and medium-height, all with their own patterns. To survive, each requires sunlight to its leaves and water to its roots. All of them take in carbon dioxide and give off oxygen, thus ensuring the life cycle of the garden's animal life. And as the leaves fall and the humus builds up, the soil becomes fertile ground for plants that are food for animals and insects, which in turn contribute to the garden. Some fertilize, some destroy noxious organisms, some aerate the soil. They work in harmony so that a continuous balance—homeostasis—is established.

Unless you are a biologist who understands the relationship of every organism to every other, the garden may seem a chaotic universe. But it is not. It is an orderly, benign cage within the confines of which its inhabitants find their equilibrium, much like the cages in which human beings find comfort and balance. My garden is a fractal, infinitely recursive and complex, the source of my fascination with the patterns that govern all our lives.

OVERVIEW

IN SIMPLEST TERMS, a pattern is an organism's
universal and predictable response to a stimulus. Some
patterns are immediate, unmediated responses to particular
stimuli: the sunflower's heliotropic pattern causes it to turn
toward the sun; the clam filters water to find its food. In
higher animals, though, patterns become more complex and
subtle, often involving multiple organ systems and minute
alterations in body chemistry. In these organisms, the neural
system may moderate and modify stimuli, routing portions of
a stimulus to various parts of the brain, which then respond
in specific ways that lead to a very complex total response.

Say, for example, you have a headache. Nothing serious, just a headache. So you take a couple of aspirin, and in half an hour or so the headache's gone. Simple, right? Not really.

This is how aspirin works.

Pain-causing cells produce cyclooxygenase, an enzyme that in turn produces prostaglandin, a chemical that tells your brain that a specific part of your body hurts. Enter aspirin. It adheres to the enzyme, causing it to stop producing prostaglandin, and as a result the pain signals don't reach the brain and you don't feel the pain anymore. What's that? You say all this is giving you another headache? But I've only just begun.

I could go on for hours.

In fact, I know a chemical engineer who devoted his whole master's thesis to an investigation of how the ingestion of two aspirin alters a human body's entire ecology. Hey, wait. Where're you going? The medicine cabinet?

In higher animals, a pattern develops when the animal's neural pathways become proficient in transporting repeated stimuli, thereby allowing the organism to respond efficiently. Patterns make it possible for the animal to carry out repetitive tasks in ways that are most energy-sparing, time-efficient, and failure-proof. Take house-training a young puppy, for instance. When he needs to relieve himself, his first instinct is decidedly not to paw politely at the door and wait for somebody to open it and let him out.

Instead, his owner must learn to recognize the signals, remain vigilant, and be willing immediately to interrupt any other activity—dinner, a favorite TV show, a romantic interlude—in order to get that puppy outside.

Although it may take a harrowing two or three weeks, the puppy will learn that outside, not inside, is the place for elimination. Repetition, in conjunction with his keen sense of smell and his natural preference for clean living quarters, has created a new pattern, one that in many dogs is so strongly established that an unexpected "accident" can be a shattering experience for them.

Millions of people, particularly those past sixty, have begun to realize that when it comes to modern communications technology, they are far behind on the learning curve. Their attitude varies from *Bah, humbug!* to a wistful acknowledgement of their ineptness to a grudging desire to learn more. This latter group includes a friend of mine who teaches English at a university. Her job now requires her to learn a new computer program. Her attempts have proved difficult and unsettling.

At first she attended several classes offered by the school. Her classmates were making good progress, she told me; they understood the instructor's directions and easily grasped the intricacies of the website. My friend, however, did not.

While the other students were happily tackling increasingly complex tasks, she was still unable even to access the site.

"I was always the smart kid," she wailed, "and now I'm at the bottom of the class."

In despair, she hired a private tutor, and finally, painfully, through tedious repetition, she has begun to learn how to navigate the new program. In other words, her neural pathways are becoming more adept at managing repeated stimuli, and a new pattern is forming.

I probably should add, though, that my friend is still feeling bruised.

"The hell with the new pattern," she tells me bitterly. "What I can't bear, what I really hate, is being flummoxed by an ignorant plastic box!"

Obviously, repeated practice grooves a more efficient patterned response. When you are first learning to drive, for example, you must attend carefully to every element of the skill, both inside and outside the car. It all feels awkward and dangerous.

But eventually, much of this learning becomes automatic: a glance into the rearview mirror, the left-turn signal, the stoplight up ahead, your foot on the brake, the wait while oncoming traffic goes by—unless you are in an unfamiliar area, this complex process, this pattern, occurs very nearly without conscious thought.

Indeed, some scientists believe that human beings create patterns to conserve energy needed for higher-order thinking. Thus, while your body and a small portion of your mind are engaged in driving, you may be thinking about the feasibility study you're going to present to your boss this morning, or devising an excuse for being late to your anniversary dinner tonight.

But a word of warning is in order: most automobile accidents occur within a mile of your home. This is because patterning creates a comfort zone, and driving a car is such a deeply patterned stimulus-response activity that if something unexpected happens, something the pattern isn't prepared to deal with (say a dog runs in front of your car), you may not be able to break out of the pattern in the millisecond required to avoid an accident.

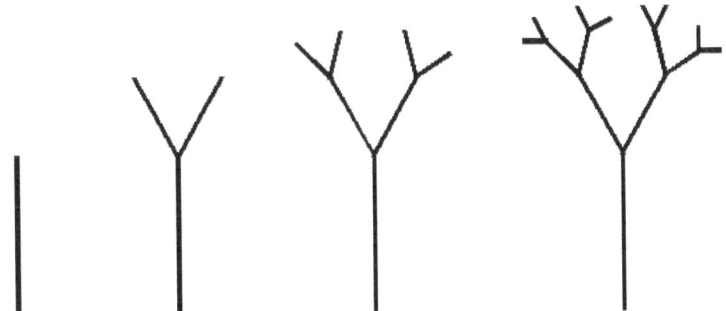

PATTERNING
REQUIREMENTS

ALL PATTERNS ARE organic in origin, and two
constructs are essential for patterning to occur.

The first is perception.

A stimulus is anything an organism can perceive, and
a response is what the organism does as a result of perceiving
the stimulus. Because an organism can't respond to

23

something it can't perceive, patterns don't develop in the absence of perception. If a stimulus is not perceived and responded to, it may cause serious consequences.

Some people, for instance, lack the ability to feel physical pain. They cannot recognize that a frying pan is too hot to touch, no stomach-ache alerts them to the possibility of appendicitis, they continue to walk on a badly sprained ankle. Such people confront daily perils that pain warns the rest of us to avoid or attend to. Even mild *mis*perception has consequences, too. One of my college professors was partially colorblind, and we students often found his wardrobe choices vastly entertaining.

If you call somebody "dumber than a rock," you mean that he is very, very stupid, because the second construct, the ability to learn and remember, is what distinguishes living tissue from non-organic matter.

Repetition makes a pattern efficient, and rapid recall helps the organism to minimize errors and maximize efficiency. Sometimes learning takes place at a simple stimulus-response level (the sunflower seeks the sun's rays; the burnt child dreads the fire), and sometimes the learning is much more sophisticated.

In humans, learned patterns can involve multiple subtle physical and psychological stimuli and nuanced responses that anticipate possible scenarios and bring about complex plans of action. A casual flirtation at the bar that evolves into a night of passionate sex exemplifies such patterns; similarly, the individuals who stop at flirtation follow different but equally subtle and nuanced patterns.

TYPES OF PATTERNS

AUTONOMIC PATTERNS

ALTHOUGH ALL ORGANISMS must "learn," that is, remember stimuli in order to pattern, learning doesn't necessarily involve thinking. Some animals, such as sponges and jellyfish, are not equipped with anything that reasonably can be called a brain; their learning is reflexive.

In higher animals, some patterns are autonomic, operating independently of the will and usually not at the conscious level. We breathe, our pupils dilate in the dark and constrict in bright light, our bodies release white cells to fight infection and produce endorphins to alleviate pain—all without our instruction. The body continuously adjusts and readjusts to maintain its equilibrium. Its genetically patterned systems respond to internal stimuli, learning and refining the patterns that support life, but we are not conscious of the patterns unless something goes wrong.

An asthma attack, for example, creates enormous distress for a person who, just before the attack happened, was giving no thought at all to his breathing. Now, as they struggle for breath, it is the only thing they think about.

Autonomic patterns that are working well provide considerable stress relief. What would life be like if we had to remember to swallow or to consciously reset our blood pressure? Not very pleasant, certainly; it's unlikely we could survive, let alone think about anything else. Thankfully, our autonomic patterning takes on these responsibilities, leaving us free to compose symphonies, write poems, provide for our families, and pursue our dreams.

PERCEPTUAL PATTERNS

BEYOND AUTONOMIC ACTIVITIES lies the realm of perceptual patterning. We see, hear, taste, feel, smell, or otherwise sense something, and that stimulus evokes a response. Given enough repetition, the response becomes etched; it acquires the status of a pattern. Perceptual patterns are protective. They keep us from being overwhelmed by stimuli. They help us filter and categorize the millions of inputs that act upon us every day, allowing us to react appropriately. We can manage the stimuli that get through, and we remain efficient and functional, dealing competently with new situations, emergencies, and decisions.

That is, most of us do, most of the time.

Some people with autism, for example, apparently have very weak perceptual filters, and they are driven nearly mad by an environment the thousands of components of which they cannot sort through and make sense of. They may devise ingenious ways to calm themselves. Temple Grandin, for example, is a woman with high-functioning autism who earned a doctorate in animal science and as of this writing, is now a professor at Colorado State University.

She has written books about autism and about animal behavior, and she is known internationally for her studies of livestock behavior, livestock facilities design, and humane slaughter methods. Grandin has noted many similarities between stimulus-processing in autism and animal behavior. Her work with livestock led her to design a squeeze-chute. The device, which stands next to her bed and which she uses whenever she is feeling over-stimulated, has the same soothing effect on her that a rancher's squeeze-chute has on a jittery horse or cow.

Another example of impaired perceptual filtering is the condition known as attention-deficit/hyperactivity disorder, or ADHD, severe cases of which adversely affect children's ability to focus, learn, and remember. Today, ADHD is one of the most commonly diagnosed psychiatric disorders in children, and treatments include drug therapy and/or behavior modification techniques. Symptoms of the disorder are often nebulous and changeable, so that diagnosis may be difficult. Unfortunately, some harassed parents, teachers, and doctors seem all too ready to apply the ADHD label to children whose real problem may be stress arising from other, unexamined sources, or boredom, or one of a host of other causes that prevent them from staying, in today's educators' grim phrase, "on task."

Recall Tony Soprano's thirteen-year-old son A.J. in HBO's *The Sopranos,* who was not doing well in school. In one episode, Mob kingpin Tony and his wife Carmela have been summoned to a conference with the school counselor, who tells them that recent psychological evaluation shows that A.J. may have ADHD.

At any rate, the boy is exhibiting several symptoms of the disorder, the most serious of which appears to be that "he fidgets." Tony considers this information for a moment. Then, never one to mince words, he offers the counselor a graphic and entirely plausible explanation of what, exactly, is most likely to be causing a pubescent boy to "fidget."

Carmela agrees.

Having effectively dismissed the possibility of ADHD in favor of their own common-sense diagnosis, the parents leave the counselor's office. (The counselor himself, we are left to suppose, probably throws up his hands in dismay at this pig-headed father and mother who have ignored his counsel.)

As for the rest of us—and A.J., who soon will quit fidgeting simply by growing out of it—we aren't likely to use squeeze-chutes to calm ourselves or drugs or behavior modification to help us concentrate . Our perceptual patterns in conjunction with our stimulus filters usually work quite well to keep us on task.

In school we may study subjects that don't interest us so that we can pass the courses and graduate. To make a living we take jobs at which we are expected to be productive even though there are elements of the work that we don't much like doing. When confronted with a disagreeable job or stuck in a tedious meeting, for example, no matter how much you'd prefer to be texting your girlfriend or playing another game of Angry Birds on your device, or even placing bets with yourself about which raindrop on the windowpane will reach the sill first, the chances are good that you'll filter out these distractions and focus on the matter at hand.

How?

By the tried-and-true method: you'll just suck it up and, as efficiently as your patterning will let you, get the thing done and over with.

As I said earlier, we pattern only what we can perceive.

What human beings perceive differs from person to person, and more importantly, from society to society. People who live in the Florida panhandle and see only a very occasional thin dusting of snow have a markedly different perception of snow from that of a native Alaskan. It is important for the Alaskan to know about snow, so his perceptual filters are wide open to it: how hard it snows, how fast it comes down, how deep it gets, its consistency, and so on are profoundly meaningful to him.

Will he be able to fish tomorrow?

Will the bush pilot flying mail and provisions from the mainland get through?

The Floridian, on the other hand, has narrow filters for snow—those few flakes, he knows, will disappear almost as soon as they contact the ground—but experience has taught him to be quite alert to wind shifts and air pressure plummets that indicate developing hurricanes and warn him to buy bottled water and start boarding up his windows.

Generally we perceive only what passes through our own filters. Hence, societies, tribes, religions, or social groups perceive events differently and build long-term perceptual patterns that are at odds. These different patterns often lead to conflict. There is no doubt that wars have been fought, feuds lengthened, and interpersonal relationships damaged because of differences in perceptual patterns between human groups.

During the Cold War, for example, Americans believed that Russian offensive weapons were trained on the United States. We saw our own weapons as defensive, but the Russian people viewed our weapons as offensive and expected that we would, at the slightest provocation, unleash them on a defenseless populace.

Both they and we lived with the threat of imminent nuclear war.

That era's doctrine of mutually assured destruction may have staved off nuclear holocaust, but it also caused considerable unease on both sides of the world. Intense rhetoric and threatening actions on the part of both superpowers over a long period of time established and strengthened their opposing perceptual patterns.

Once a pattern is created, it is stored.

It may be used every day or never again, but it is always there, ready to be activated by the proper stimulus. If you learned to ride a bike as a child and your muscular and cerebral systems remain intact, you'll be able to ride a bike when you're eighty, even if you haven't climbed onto one since you were twelve. The very complex patterns of balance and control of a bike have lain dormant all those years, but are immediately ready to go to work when the stimulus calls. You may be a little unsteady at first, but within minutes you'll be riding around the neighborhood as freely as you did as a child.

The same is true of societal patterns.

If a truce is declared in a conflict, but something disturbs a long-term cessation of hostilities, the old patterns may come to the fore, resulting in renewed fighting. The seemingly endless cycle of war and peace in the Middle East is an example. In that area today, protesters in country after country are rising against their dictatorial heads of state, clamoring for liberty.

Someday, perhaps many years from now, the people will achieve their aims, good governance will take root, and peace will reign among the Arab nations. Such a future can happen, however, only if the Middle East's multitude of tribes and factions renounce their old patterns of mistrust and enmity and try consciously to create new, mutually beneficial patterns and habits of mind.

ASSOCIATIVE PATTERNS

ASSOCIATIVE PATTERNS ARE a subset of perceptual patterns. They are in some sense *conditioned* responses. Pulling your hand away from a hot stove is an *unconditioned response*—instinctual, automatic, and not under conscious control. It's a reflex, and it's lightning fast because it doesn't have to involve the brain. The stimulus travels through a short reflex arc made up of nerve fibers, and the response is immediate. It had better be, or you'll burn your hand. Conditioned responses and their close corollaries, associative patterns, are learned responses that are very similar or identical to unconditioned responses, but they are produced by a conditioned stimulus. Ivan Pavlov, of course, conducted the classic work in conditioned response.

Pavlov, who was studying the mechanics of digestion, knew that dogs salivated when presented with food; salivation was the dogs' unconditioned response to the food stimulus. Pavlov noticed, however, that his laboratory dogs drooled even in the absence of food.

Through observation, he discovered that his dogs were responding to a conditioned stimulus—the lab coats worn by those who fed them. Every time the dogs were fed, the person bringing the food wore a lab coat; the dogs had learned to associate the lab coats with food, so they drooled whenever they saw a person in a lab coat, regardless of whether he was carrying food.

Associative patterns occur in humans as well. One such pattern is what doctors call "white coat syndrome," a phenomenon almost comically Pavlovian. The condition occurs among some people who have had frightening experiences with doctors. In a hospital or doctor's office, the appearance of a doctor wearing a white lab coat causes these people considerable distress, whether or not they are actually sick patients. Their blood pressure rises, their pulse races, their respiration rate increases, they may start to sweat, they become tense. It isn't the doctor himself who is responsible, but the white coat he is wearing, which the uneasy person has learned from previous experience to associate with a frightening outcome.

Sometimes associative patterns have unexpected results. Because we are animals with memory, an associative pattern may grow out of retained stimuli. That is, we can remember a stimulus, even one as ephemeral as a dream, and we can remember the response it caused. Then, when we think of the stimulus, our mind and body may respond as if the stimulus were being externally applied and directly perceived at the present moment.

Psychotherapists, for example, know that age-regression through hypnosis can be a thorny procedure. A woman under hypnosis may recall an instance of childhood sexual abuse in such vivid and horrific detail that she is convinced it is happening again, *right now*, and she is in effect re-traumatized—the very antithesis of what skilled psychotherapy should achieve.

Or consider John and Mary, who are planning a back-yard garden. John is a real enthusiast; he talks about the pleasures of working the soil, planting the seeds, and observing the life of garden insects. Mary nods in agreement—until the part about insects, whereupon she recoils in fear, and John is aghast.

Was it something he said?

Yes.

It turns out that Mary was once stung by a bee as she was weeding a garden. She had nearly died in anaphylactic shock, and from that time to this, she has dreaded even thinking about insect life. In this instance, the perception that Mary receives—the sound of John's voice talking about insects—stirs an existing chain of stimuli (buzzing, stinging, gasping, choking, unconsciousness) and causes her to respond not only to their current conversation, but also to the latent stimulus, the near-fatal sting. Thus a minor stimulus sometimes can cause a startlingly disproportionate reaction.

Associative patterns can have roots that extend into prehistory. An unhappy result, for example, of one of these ancient associative patterns is that the United States has become one of the fattest countries on earth. Overeating is a national pattern. Although we of course must eat to survive, during almost all of the millions of years of human history, survival was the primary reason to eat, for famine was always lurking.

We were wanderers pursuing food—herbs, roots, vegetables, fruits, legumes, and meat when we could kill it. When food wasn't available, as in a drought, we starved; when it became available again, we gorged. In temperate climates food was scarce in winter, so that the buildup of fat in summer helped ensure survival in cold weather. Like other carnivores, evolved a pattern of eating more than enough to assuage hunger when food was plentiful.

An abundance of food, moreover, was a happy time—something to celebrate. It is not by accident that when a joyful event takes place, a wedding or a bar mitzvah, we still celebrate with a feast, and at a feast we usually eat too much because it's...well, because it's a feast. Through the millennia of our history, overeating became associated with both survival and happiness.

For some people, patterns of modern civilization have superseded the old seasonal patterns of feast or famine, but the associative patterns of overeating have not changed. Generations ago, fish from the Gila River, plus a few fruits and vegetables, made up the principal diet of the Tohono O'odham tribe of Arizona's Sonora Desert. But as southern Arizona's urban population increased, irrigation projects and municipal water diversions dried up the Gila; it is no longer a source of food for the Tohono O'odham.

But "civilization" brought new sources, and to these tribal people, most of whom live in poverty, the cheaper the sources were, the better. Pepsi, potato chips, hot dogs, refined flour and sugar products, all produced with liberal amounts of trans-fats and sodium, were staples. Fishing and farming were out; indolence was in. So was eating. The result was that the formerly lean and healthy Tohono O'odham became known as the fattest people in the world. And unsurprisingly, their once minimal rates of diabetes and heart disease skyrocketed.

Today, most Americans have an overabundance of food in all seasons. Our pantries and refrigerators and freezers are bursting with an amount and variety of food that even our grandparents, much less our ancient ancestors, would have found astonishing.

But do we consume that huge store of provisions before we buy more?

Of course not.

Maybe nothing we have on hand seems particularly appetizing for tonight's dinner—we feel like having lamb chops—so it's off to the supermarket to get some, and then a quick stop by the bakery for one of their great sourdough loaves, and those pastries look awfully good too, and then let's just pop into the wine shop because we can't remember whether there's any pinot grigio left in the wine rack.

In a persistent pattern that has not changed much since prehistory, we overeat when we have food (which is all the time), and we overeat to make ourselves happy (even if we're already happy). It's an associative pattern squared.

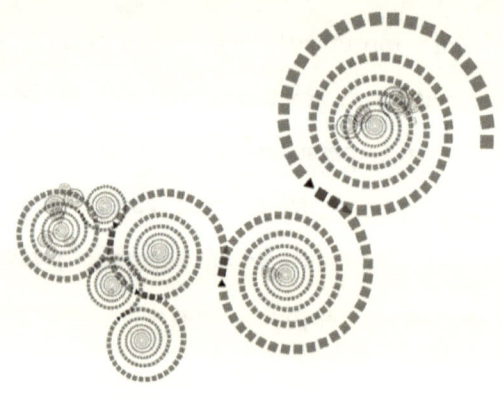

HOW PATTERNS
CHANGE

BECAUSE THERE ARE no entirely new living tissues to observe, at present we are not able to study an entirely unpatterned cell. This makes it hard to understand the fundamental origin and development of patterns. By

42

studying organisms as they change, however, we can learn a great deal about how patterning occurs at the cellular level. One thing we know is that pattern change is usually very difficult. Patterns are deeply ingrained, so that it takes tremendous internal and external pressure to alter them.

Nonetheless, significant environmental change may have a profound effect on the organism, complicating its stimulus-response pattern and additionally burdening an already stressed and disordered entity. For a time the organism may employ old strategies that are no longer valid or necessary in its new circumstances, thereby consuming energy that would otherwise help it survive. In these circumstances, many organisms die off. Some survive, though, and move again toward energy conservation and efficiency.

A dramatic case in point is the history of medical treatment for staphylococcus infections. These infections can be especially opportunistic when the body is under stress, so it is not uncommon for patients in a hospital, whose immune systems are already compromised, to develop a staph infection.

Like other organisms, staphylococcus bacteria use their energy to keep themselves alive and reproducing. In the past, these bacteria won the war with human beings because there was nothing to stop them, that is, nothing to interfere with their homeostatic functioning. Although most people attacked by the bacterium were able to conquer it and build antibodies against it, some died.

Among soldiers in World War II—and probably in every other war, for that matter—staph infections were common precursors to gangrene and amputation and, often, death. Perhaps you remember Harry Street, the writer in Hemingway's "The Snows of Kilimanjaro" (the short story, not the movie) who lies dying of gangrene. A thorn puncture to his leg caused the initial infection, and I have always thought that it was most likely a staph infection. Staphylococcus respected no one, not even fictional characters.

But the advent of antibiotics changed all that.

First came sulfa drugs. Sulfa altered the environment in which the bacteria thrived, disturbing their balance so radically that most of them could not deal with the change; they died, and people recovered. Some of the bacteria, however, adapted and survived. They did so with a combination of two mechanisms. One is simple survival of the fittest.

The second is the creation of mutant species, genetically equipped to respond efficiently to the new environment. These adaptive bacteria reproduced, creating a sulfa-resistant strain that not only survived but counterattacked. The tide turned, and bacteria once again had the upper hand.

In 1928, when Sir Alexander Fleming discovered penicillin, the scales tipped in favor of humankind. In the early days of penicillin's use, small dosages of 15,000 units wiped out almost all staphylococci, but some grew resistant to this drug too, spawning billions of descendants, so that penicillin became increasingly less effective against them.

The dosages required to defeat them rose from 15,000 units to more than a million units every four hours; today, many antibiotics that used to be considered wonder drugs have virtually no effect, and people continue to be at risk from staph infections.

Had the environment continued to change faster than the bacteria could adapt, their numbers would have diminished to the point that they would have stopped replicating. But it takes time to develop and test antibiotics, and thus far staphylococcus has kept pace with the introduction of pharmaceutical threats to their survival. And so the war goes on. One of the more recent permutations of staphylococcus is methicillin-resistant staphylococcus aureus, or MRSA. This dreaded scourge of hospitals is immune to most antibiotics, and though a few can usually treat it successfully, there is evidence that they also are becoming ineffective.

Just as it's difficult for a simpler organism to change its patterns, it's hard for humans, too. Take for example our centuries-long romance with tobacco, which only in the past fifty years has begun to turn sour. I used the word "romance" advisedly. Some time ago I attended an evening of Renaissance music; "Tobacco" was sung as a madrigal. It was quite beautiful:

> *Tobacco, tobacco,*
> > *Sing sweetly for tobacco!*
> *Tobacco is like love, o love it,*
> *For you see I will prove it.*
>
> *Love maketh lean the fat man's tumour,*

So doth tobacco.
Love still dries up the wanton humour,
So doth tobacco.

Love makes men sail from shore to shore,
So doth tobacco.
'Tis fond love often makes men poor,
So doth tobacco.

Love makes men scorn all coward fears,
So doth tobacco.
Love often sets men by the ears,
So doth tobacco.

Tobacco, tobacco,
Sing sweetly for tobacco!
Tobacco is like love, o love it.
For you see I have proved it.

I picture the dashing explorer Sir Walter Raleigh, his velvet cape flung glamorously about his shoulders, being serenaded at Queen Elizabeth's court for his prowess in bringing, among other excitements from the New World, that pleasing new stimulant, tobacco, home to England.

Three hundred years later, here in the New World, men were still extolling the virtues of tobacco. Advertising men. The Marlboro Man made smoking Marlboro cigarettes synonymous with everything the American male longed to be:

47

tall, handsome, strong, independent, sexy—a real force of nature.

Actually there were several men, a half-dozen or so, who acted the part in the Marlboro commercials. Three of them died of lung cancer. Cigarettes were a key element in the public image of other icons of the twentieth century. Rod Serling, for example, a prolific writer probably best known for his *Twilight Zone* television series, was a four- or-five-pack-a-day smoker for most of his life. Serling developed heart disease.

During open-heart surgery, he suffered a final, fatal heart attack. He was fifty. The actor Humphrey Bogart, famous for his tough-but-tender-guy movie roles, also smoked heavily for many years. Bogart died of esophageal cancer. He was fifty-seven. The brilliant broadcast journalist Edward R. Murrow chain-smoked Camels. He died of lung cancer. He was fifty-seven.

A Maryland woman I know remembers from the forties the annual Tobacco Pageant at the Charles County Fair. Tobacco was Southern Maryland's main cash crop (it made some people very rich), and the little outdoor drama showed the centuries-long connections between tobacco farming and the area's prosperity.

There was also a Tobacco Queen, my friend recalls, a winsome local high school girl. Now, before the tobacco plant is cut and its leaves stripped from its stalk and hung to cure and turn brown, it is a lovely thing, its leaves long, deep-green and velvety. During the coronation—the pageant's high point—when the high school principal crowned the Queen, she then was clad in a royal cloak made of green tobacco leaves sewn together with thick golden thread and closed at the throat with tasseled golden cords. What an honor it was. And the name of the new monarch?

Queen Nicotina.

With smoking widely perceived as a glamorous, sexy, even healthful pastime, and with very few warning voices raised against it, its popularity among the general public was secure.

And for a very long time it was equally popular among many doctors, whose insight into its dangers seemed about on par with that sixteenth-century madrigal composer's. You may wonder just how much medical acumen is necessary for a doctor to adduce that sucking smoke and its attendant noxious chemicals into your lungs two or three hundred times a day is hazardous to your health, and the answer is, not much.

But doctors can be as averse to change as anyone else, particularly if they themselves are smokers. So it was not unusual for a doctor actually to advise a patient to smoke. Feeling a bit nervous? Here, have a cigarette. Scratchy throat? Persistent cough? Here, have another. It'll soothe your throat and quiet that cough. Most discreditably of all, a few doctors went so far as to sell their reputations to advertising companies, appearing in print ads praising the benefits of smoking.

By the late fifties, however, medical opinion had begun a massive shift. In 1964 the results of a two-year study of the health effects of smoking were published in Surgeon-General Luther L. Terry's widely publicized *Report on Smoking and Health*, and by 1968, 78 percent of Americans had come to believe that smoking caused cancer. The logical next step, then, would be that everybody would quit smoking, right? Wrong. In a contest between logic and a deeply established pattern, the pattern is the stronger contender.

We know that smoking harms us, that it may kill us. But if we've had a cigarette with our morning coffee every day for fifteen years, then merely the aroma of coffee will evoke the smoking response. If we smoke when we're stressed, and stress arises many times a day, we're going to want a pack of cigarettes every day. Following the pattern creates a feeling of contentment and release—a sense of restored equilibrium—even though we know that the activity itself threatens us.

That is why behavior modification plans, such as smoking cessation programs, often fail. During the time in which we are trying to change the pattern, we attend to it. We think consciously about it, and we may have some short-term success in modifying our behavior by increasing our awareness of what stimulates it and blocking or rerouting the response.

Instead of having a cigarette, we may substitute a piece of hard candy. Once we stop focusing on the goal, however, the old non-cognitive stimulus-response patterns reassert themselves. For a behavior-modification program to work, it must alter not only the behavior itself, but also the underlying perceptual pattern, and it must block or redirect the stimulus.

In the case of smoking, a great number of stimuli impel the smoker to light up: a traffic accident narrowly averted, an argument with his wife, the eleven-o'clock news sign-off, winning (or losing, or playing) a Scrabble game, a glass of wine, seeing somebody else light a cigarette, debarking from a long airplane trip—just about anything can serve as an excuse, or stimulus, to have a cigarette.

A geologist friend of mine, a heavy smoker for thirty years, spends every December, January, and February at a small campsite in Antarctica, where he and a little group of like-minded scientists extract ice-core samples as part of their study of climate change. He does not smoke at all during those three months. I asked him why. He isn't sure, he tells me, but "everything seems so clean. The air is so pure. I just don't like to muck it all up with cigarette smoke."

My friend cares deeply about the future of our planet, and I'm sure that what he says is true, but I don't think it is the underlying reason that he doesn't smoke in Antarctica. I think the environment—free of traffic and television and disagreeable people and so many other stimuli that he is accustomed to at home—is the reason.

Antarctica is also nearly free of color; you see mostly white, so that the snow- and ice-scape presents, almost literally, a blank slate upon which it is easy to impose a new pattern, e.g. not smoking. And yet, despite three months of cigarette-abstinence, despite feeling healthier and enjoying renewed vigor, despite all that logic and common sense are screaming into his ear, as soon as he arrives home my friend starts smoking again.

Why?

"Oh, the traffic from the airport was bad. The news on TV is terrible. I wanted a drink but somebody had broken in and stolen my liquor, and one of the water pipes froze and cracked while I was gone, and, you know, it was just…well, everything." This time he was on the money. It *was* "everything"—all those old familiar stimuli that produced the old familiar response that he knew would make him feel better, no matter how briefly. It was impossible for him to resist those innocuous-looking little white cylinders. So round. So firm. So fully packed.

So free and easy on the draw.

PERCEPTUAL PATTERNS IN LIVING

AS A PERSON grows through maturity, various functions begin to change or diminish, and those changes affect the stimulus-response grooves. With age, for example, the sense of taste becomes somewhat vitiated; as this happens, your food intake may correspondingly increase, in a vain attempt to recapture the flavors you remember by eating more, or decrease, because food just doesn't taste the way it used to.

You also may find sharper-flavored foods more satisfying than you once did, because they stimulate your time-dulled taste buds.

Or perhaps you've begun installing higher-wattage light bulbs in your house. No, it isn't those new CFL bulbs that are the problem; it's your aging eyes.

To a great extent, however, patterns that were etched early in life and have been in use for a long time control our actions. The rigidity of these patterns can obstruct our efforts to respond to new stimuli and develop new patterns, so that we prefer to cling to experience we know, to things as they have always been; we attempt to fit all external stimuli into a limited number of input slots.

If your aging computer has ever run out of memory, you actually can see how hard it tries to incorporate new data, and how awkward its efforts seem. It is a pitiful thing to observe, rather like a very old person, his mind clouded by age, trying to remember all his grandchildren's names and ages, and which of his children they belong to. Irrational and inappropriate reactions often occur when new stimuli enter an environment no longer equipped to deal with them. Your old computer may crash; the old person may cry.

The ability to keep perceptual entrances open and, indeed, to create new entrances, probably is not a function of intelligence (though it may be a matter of brain function), but a mechanism of self-examination that allows us to challenge our own perceptions and, if we find them no longer useful or even false, replace them with new ones. People with a highly developed ability to challenge patterns, who habitually question whether what they are perceiving is in reality different from their perception of it, are able relatively easily to change.

As long as their minds work well, such people lead dynamic intellectual lives even when their bodies have become infirm. Think for example of the theoretical physicist and cosmologist Stephen Hawking, who was diagnosed at age twenty-one with a motor neurone disease, a sort of atypical amyotrophic lateral sclerosis (Lou Gehrig's Disease).

At the time of his diagnosis Hawking was given three years to live; he is now in his seventies. The disease has progressed to the point that he is almost completely paralyzed, as well as suffering with a host of other severe physical debilities. Nevertheless he continues to make important contributions to science. Hawking's goal is "complete understanding of the universe, why it is as it is and why it exists at all." Now, there's a goal that *demands* challenging our own perceptions.

INDIVIDUAL PATTERNING

IT IS IMPORTANT to understand that no two organisms—paramecia, dogs, humans—are patterned in exactly the same way; each living organism is a combination of genetic template and environmental experience.

It is the old nature versus nurture argument, but in fact there is no reason to argue. As research psychologist Gary Marcus puts it in *The Birth of the Mind*, our brains are "*prewired*—flexible and subject to change—rather than *hardwired*, fixed and immutable." Genes are critical to wiring the brain, he writes, but the environment is involved in "building a person." Because every organism's genes and experience are unique, every organism's patterns are unique. Although we can generalize about group patterning by observing only a few members of the group, we can't discern each member's specific patterns. And the more complex the organism, the more variety in its patterns.

Cardinals nesting in a garden are an example. These birds have patterns of nest-building and baby-tending common to all cardinals and recognizable among birds in general. But the patterns of a pair of cardinals nesting in one garden will differ from those of a pair in another garden because the physical environment is different, which impacts the availability of materials for shelter and food.

Perhaps it is a question of building materials: one pair finds some nice dog hair to line its nest and the other uses soft leaves. Or a cat lives in one of the gardens, and those cardinals, wise through experience to the bird-stalking patterns of cats, build their nest higher than the other pair's nest.

Cardinals are highly complex creatures, but in comparison, what a piece of work is man, how noble in reason, how infinite in faculties.

Among humans, the protean interplay of genes and experience is what keeps patterning from turning us into automatons. Genetic codes usually carry some potential for novelty and creativity, and if these genetic gifts are nurtured in a rewarding environment, the patterns that develop can be highly creative. Moreover, as the mind amuses itself seeking novelty, it may detect subtle connections among experiences and create exciting new ideas. It is doubtful, for instance, that Vincent van Gogh's genius could have come to flower without the lifelong emotional and financial support of his brother Theo. And where would we all be if that apple hadn't dropped onto Sir Isaac Newton's head? Still falling off the earth and drifting around in the sky, that's where.

It now appears certain that there are strong genetic components in disease. This comes as no great surprise; we all know of illnesses, or predispositions to illnesses, that run in families. If Bob's father and grandfather died young of heart disease, then Bob too is likely to die young of heart disease. If Bob is smart and well-informed, however, he will jettison the patterns of activity, such as smoking and overeating, that exacerbate the predisposition, and replace them with new patterns—daily exercise and a moderate diet—and thereby can hope to avoid early death.

We all are generally aware, if only vaguely, of the role heredity plays, but now science is nailing down the details. Medicine itself is being redefined to identify disease by its genetic involvement; our entire understanding of disease is being recast.

My niece developed breast cancer, a disease that had proved fatal to both her mother and grandmother. Tests revealed that my niece harbored one of the BRCA gene mutations that cause her specific type of breast cancer. Her three daughters, all around forty, were also tested, and one of them carried the same mutation.

That daughter immediately elected to undergo a prophylactic bilateral mastectomy and oophorectomy (ovary removal); having this surgery significantly increases the odds that she will not incur that particular cancer. More and more women today are electing to have the same surgery for the same reason. Take Angelina Jolie, for example. For some of them it may be a wrenching choice, but most of them recognize it is the wise one.

I admire their intelligence and their fortitude. In this society, where our patterning tells us that women's breasts are objects of beauty and lust, a *sine qua non* of motherhood and womanliness, a woman's breasts are often so profoundly a part of her physical and psychological self that their removal means she must seek new strength within herself, and new patterns, to compensate for her loss.

PATTERNS IN SOCIETY

IN VIRTUALLY ALL circumstances, perceptual patterns affect human beings' responses, yet we are largely unaware that our responses are patterned. We may believe that our reactions arise *de novo*, but that is almost never true.

As we grow from infancy toward adulthood, we learn that various behaviors result in rewards or punishments, so we develop patterns that win the reward or avoid the punishment. Early patterns persist throughout our lives, affecting all our relationships. The patterns may exist below our level of cognition, but they are nonetheless strong, rigid obstacles to our recognition of competing stimuli.

How frequently is it the case, for instance, that men marry women much like their mothers and women marry men like their fathers? The patterns instilled by children's earliest love-object, their mother or father, are permanent, and though when a child reaches adulthood he may no longer be conscious of them, those patterns influence his choice of a spouse.

Our perceptions are influenced by our environment, our family, our community, our society. Most Americans, for example, think that the smell of steak sizzling on an outdoor grill is quite appetizing, but in cultures where the cow is sacred, the smell can be nauseating as well as damaging to a person's sense of spiritual well-being.

Similarly, some dietary preferences of other cultures are incomprehensible to Americans. Kimchi is a staple for Koreans, but many Americans often find its odor repellent. And roast dog? Don't even think about it. Back here at home, many men affect to love raw oysters; at cocktail parties they make a great show of slurping the slimy creatures from the half-shell. That is because oysters have been known for thousands of years as an aphrodisiac.

(Their high zinc content may stimulate testosterone production, but it takes an awful lot of oysters actually to make a noticeable impact on potency.) Oystermen, on the other hand, the men who ply the trade of oyster-fishing, will eat them fried, frittered, stewed, baked in a pie, and Rockefellered; they will eat them, in fact, any way except raw.

Perhaps oystermen believe that their work relationship with these unattractive mollusks is plenty intimate enough. Perhaps they choose to avoid the lethal bacteria that may be lurking in raw oysters. Or perhaps, unlike those fellows at the cocktail parties, these tough, intrepid watermen are confident they have nothing to prove.

When we band together in families and tribes and cultures, we bring our patterns with us. We become the "genes," or templates, in the new community, and throughout our lives we act on the community and it acts on us. Some of the genes are dominant and some are recessive. As perceptions enter the communal body and are modified by the "genetics" of the group, cultural patterns form and are codified as "the way things are" for that community.

Once established, the patterns go underground into the non-cognitive sphere. Thus an entire community may hold a set of patterned beliefs and yet, unless they are challenged, be largely unaware of them. Some of the Amish sects illustrate this point. These sects place great pressure on their members to conform to their beliefs and strictures and thus to maintain the status quo. Children are not educated past the eighth grade, as it is thought that additional schooling isn't necessary to a successful Amish life, and indeed that further education may implant new ideas dangerous to the community's equilibrium.

These attitudes stultify independent thinking and creativity, but seemingly few members of the sects actually *feel* stultified, because it is the way things are.

Much like autonomic and sensory patterns, communal patterns protect us from danger, ostracism, exile, and even death. Yet the perceptions on which the patterns depend vary widely from family to family and society to society. A young white child, steeped in the culture of the Ku Klux Klan, will obviously have far different perceptions of African-Americans and Jews than those of a child brought up to believe in the dignity and worth of every person.

Barring the intercession of conscious thought, each child's patterns related to race and culture will remain fixed, for social patterns take on a life of their own. They live in the minds, the actions, and the perceptions of the group members and can program the neuronal structures of successive generations. The patterns may remain in place long after they have value for the person or for the group itself; once established, they are very hard to change or eradicate because like all organisms societies too seek homeostasis.

Short of war, famine, or some other catastrophe, modifying a social pattern takes great energy and a very long time. In parts of America, segregation had been the accepted social pattern since slavery ended in the mid-nineteenth century. Then, in 1954, the Supreme Court put an end to segregation in public schools. A decade later the Civil Rights Act became law. Laws change behavior, but are much less effective at changing attitudes, as evidenced by the protracted unwillingness of so very many white Americans to challenge their patterns of thinking about black people even as the law forced them to change their actions. And it is clear that far too many people continue to cling to those outworn and socially damaging patterns, even if it's not about skin color.

However, in a complex society, if the pressure is strong enough, perceptual shifts of great magnitude can happen quickly. Earlier I mentioned the plight of Japanese-Americans during World War II. Before the Japanese attack on Pearl Harbor, they generally were considered good citizens, working diligently alongside the rest of us to achieve the American dream. But after the attack, in which they had no part, they became "Japs"—our enemies. The fervor engendered by the events at Pearl Harbor easily provided enough pressure to change the country's perceptions of Japanese-Americans almost overnight. In mid-February 1942, President Roosevelt signed Executive Order 9066, which called for the relocation of 120,000 Japanese-American citizens and legal residents to internment camps for the duration of the War. We no longer perceived the pejoratively named "Japs" as citizens of our country.

Other circumstances may have been partially responsible for the President's quick and dramatic action. For one thing, before the Pearl Harbor attack many Americans already were angry about Japanese competition with American interests. For another, most of the first Japanese immigrants arrived in this country to work on railroad construction; their immigration was tightly controlled and they were subject to oppressive rules.

Moreover, even after the immigrants became American citizens, they and their American children and grandchildren were often set apart by their customs, language, and looks. Thus we could intern them and appropriate their property with equanimity because "they had always been different." This previous pattern of thinking reemerged after Pearl Harbor and no doubt increased the hostility. Fortunately, after the war and a period of adjustment during which American society developed new patterns, by the late twentieth century Japanese-Americans had become fully respected citizens of the United States. It didn't hurt, of course, that by then the generations that remembered the war most vividly were dying off, and that the younger people who replaced them would be more accepting of differences, more willing to ignore ancient enmities, and more ready to form new patterns attuned to new circumstances.

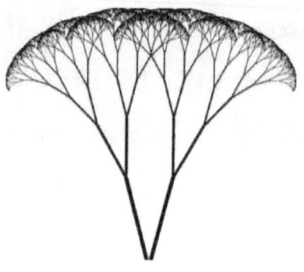

PATTERN DEMANDS
OF THE
NEW SOCIETY

IT IS A truism that change accelerates in time. But today many of us are asking plaintively, Does it have to accelerate *so darn fast?*

We are caught in a maelstrom of information as a result of technology. Only with careful, painstaking thought can we sift through the masses of information pouring in from myriad sources and successfully distinguish fact from fiction: what is so from what somebody just wishes were so. This rapidly expanding universe of information is causing problems for people and for societies; for both, patterns are well-established and long-term. When we find ourselves in an environment in which changes are happening fast and we must therefore modify our perceptions and beliefs quickly, conflict arises between our natural preference for stability and the alarming tempo of change.

Patterns are hard to shift, and people need time to adjust their perceptions. But time is a luxury no longer available to us. An enormous asteroid from cyberspace is hurtling toward us; we can't see or hear or touch it, but we know that it's coming and we're already feeling its effects. The rapid advances in information technology are creating vast changes in every field of human endeavor: education, science, medicine, communication, the arts, engineering, law, agriculture....

I can think of almost nothing in our everyday existence that has not been affected by information technology. And those of us who can't or won't keep pace with change, who ignore or deny or make light of that asteroid, will suffer the fate of the dinosaurs. Oh, it's not a real asteroid that destroys outright. It's worse. If we allow technology to leave us in the dust we'll endure a kind of living death as, at worst, nonfunctional units of a dynamic society, and at best as harmless anachronisms.

The same is true for societies: the magnitude of the changes we now confront has altered the rules so that the risks of refusing to change are often more dangerous than the risks of change itself. To meet the challenge, people and the societies they inform must redefine stability to include knowing how to readjust quickly, to recognize and respond to new perceptions, and to form new patterns faster.

Consider for example turbulence in the Middle East. President Hosni Mubarak of Egypt has been toppled from his dictatorship by bright, technology-savvy young people insistent on social and political change. In Libya, Moammar Gaddafi's long and oppressive reign has been destroyed—and Gaddafi himself killed—by armed rebels and NATO forces wanting similar change. The patterns of thought and behavior of both those men didn't permit them to acknowledge that though once there was a "time to hold 'em,' *now* was the "time to fold 'em."

There is turmoil in Syria, uprising in Bahrain. Unrest appears even in Saudi Arabia, that most obdurate of theocracies.

It is unclear how events in the Middle East will finally turn out. One certainty, though, is that as new leaders assume power, they and their governments must deal nimbly with change, or they too will go the way of Mubarak and Gaddafi. Hidebound, mired-in-the-past governance has no place among people and nations whose longing for change has been aroused.

Societies that cannot re-pattern quickly and effectively are destined to falter in their relationships with societies that can, to sink into inconsequence, perhaps to disappear altogether.

Societies that will prevail are the ones prepared and ready to play meaningful roles in our swirling, dizzying, ever-changing new world.

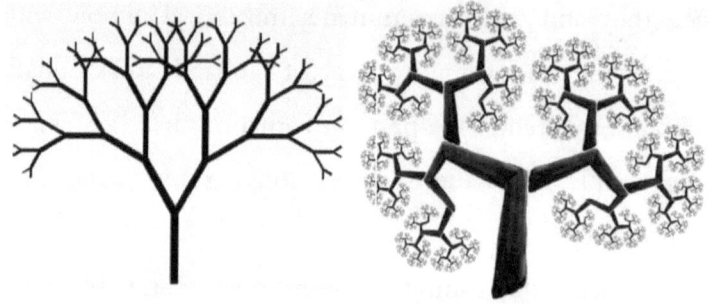

PRE-BIRTH
PATTERNS

IN 1695, THE Dutch mathematician and physicist Nicolaas Hartsoeker conceived and drew an image of tiny, alien-looking men-creatures he called homunculi. A follower of a scientific school known as "spermists," Hartsoeker postulated that these tiny men were curled up inside each spermatozoon, which would be placed inside a woman for growth into a child. The image he drew has been reproduced for centuries in history, science, and medical textbooks. Since Hartsoeker's time, Western science has made advances in the study of how babies develop in the womb, but it has taken a while. There were physical and psychological obstacles to be overcome.

First, there was the fact that nobody could actually see what was happening in utero. Museums and medical schools might be well supplied with fetuses (the result of miscarriages), usually preserved in glass jars, at various stages of development, and indeed these specimens, as well as other artifacts of pregnancy, offered a rich source of empirical information.

But until the second half of the past century, how a fetus spent its day remained a mystery. Did it hear? Feel? Play? Experience emotion? Could it, possibly, think? Nobody knew for sure, and science's default position was to focus on the finished product: the successfully birthed live baby itself.

For many centuries, most people considered sexual intercourse, pregnancy, and childbirth a perfectly natural sequence of events and an appropriate topic of private and public comment. English lords and commoners alike, for instance, hoping for a boy, closely observed Anne Boleyn's first pregnancy by Henry VIII. They were disappointed, as the baby was a girl, but Boleyn's second pregnancy aroused their hopes again, and their even keener interest.

All of London correctly calculated almost to the instant when she conceived, no detail of her condition escaped the people's notice, and they learned of her miscarriage (it was a boy) within a day of its occurrence.

Queen Victoria reigned from 1837 to 1901. Her immense influence on her own subjects extended into America and lasted well into the twentieth century. Queen Victoria herself has been blamed unfairly for a lot of silliness. For example, the story that, on her daughter's wedding night, the Queen advised her to "lie back and think of England" as a means of coping with her bridegroom's demands, is most probably false. Scholarship has revealed that Victoria and her beloved Prince Albert enjoyed a robust sexual life together. Nevertheless, "Victorian" defined many aspects of that era— fashions, furnishings, architecture, literature, politics, societal attitudes. Particularly among the middle or aspiring-to-be middle classes, it became an article of faith that one must "observe the proprieties," a thick social veneer of written and unwritten imperatives that must have made those who obeyed them seem as stuffy and airless as their Victorian parlors.

Childbearing was seen as a patriotic duty—Victoria bore nine children—yet women were assumed to take no pleasure in sexual intercourse, and if they did, then most certainly they must never admit it. In polite society, words describing natural functions of the body became unsayable; peculiar euphemisms took their place.

A pregnant woman was said to be "in a delicate condition." Childbirth itself was "confinement." And as the pregnant woman herself began increasingly to "show," she curtailed her activities and social engagements more and more, so as not, I guess, to offend others' sensibilities, or advertise a condition that could have come about only through—horrors!—intercourse. Toward the end of her pregnancy the woman became somewhat of a prisoner in her own home.

In 1938 *LIFE* magazine published a pictorial, "The Birth of a Baby," consisting of some thirty-five black and white slides taken from a film of the same name. Nine of those pictures caused an outcry, not so much among the majority of the public as among a good many politicians, clergymen, and civic leaders—the usual crowd who mind other people's business.

They called the pictorial outrageous, immoral, salacious; they predicted it would destroy the morality of American youth. Thirty-three U.S. cities, plus Canada and Pennsylvania, banned the magazine, sending police to confiscate all copies from the newsstands. *LIFE*'s publisher, Roy Larsen, was arrested and charged with indecency.

The offending nine slides purport to show a baby's birth. In the first of them, a baby's head is emerging from its mother's vagina, but without the caption that explains what is happening, you really can't be sure. Every part of the mother's body is completely covered in sheets or towels, so that the baby appears to be making its way out of a pile of laundry. The other eight pictures show standard postnatal procedures: the newborn is tidied, weighed, measured, swaddled; its eyes are treated to avoid infection; it is presented to its happy mother. That's all. Those are the salacious pictures.

Slowly attitudes lightened up, however, and patterns began to change. Pregnant women got out of the house. Boldly they drove, they worked, they attended social events.

Obstetrics became a specialty and "maternity clothing" became an industry. Those smock-like blouses with ruffles and puffy sleeves apparently were designed to camouflage a woman's expanding abdomen and to suggest that, really, she was still an innocent child.

The design failed, on both counts.

Eventually, in public as well as in private, pregnancy became a condition to celebrate. Still, however, the life of the fetus remained a puzzle, and to many pregnant women the fetus itself, unseen and unheard, didn't seem exactly "real" until birth magically transformed it into a baby. Doctors didn't help much.

Even though it was clear that whatever a pregnant woman ingested would affect the fetus, pregnant women drank alcohol and smoked cigarettes, frequently with their doctors' blessing. It wasn't a matter of ignorance; the problem was that habit is so often an obstacle to common sense, and that it was hard to consider the hidden fetus a genuinely human creature. But then *LIFE* entered the picture again.

Its April 30, 1965 cover featured a full-color photograph by Lennart Nilsson of an eighteen-week-old embryo *inside* its amniotic sac, and the accompanying sixteen-page article included more of Nilsson's extraordinary endoscopic in-utero photography, showing various stages of fetal development. In one picture the fetus is sucking its thumb. The time was ripe: it was the Sixties, most of the "proprieties" were falling into desuetude, *LIFE* sold eight million copies in four days, no one was arrested.

In ensuing years, many other publications followed *LIFE*'s lead, presenting similar displays of in-the-womb photography. In the Eighties, ultrasound scanning became a commonplace of obstetrical practice; today parents can opt for 3-D and even 4-D ultrasound pictures. Along with continuing advances in photographic techniques has come a torrent of new information that helps ensure the health and well-being of both mother and baby. And now that the fetus's privacy has been breached and the heretofore unseeable can be seen, a woman *knows* that she is incubating a genuine, one-of-a-kind human child, and that she needs to nurture it throughout her pregnancy. No more coy euphemisms. No more frilly smocks. In fact, today's woman is likely to choose clothing to emphasize her pregnancy rather than disguise it.   Where pregnancy and childbirth are concerned, ours is an age of enlightenment.

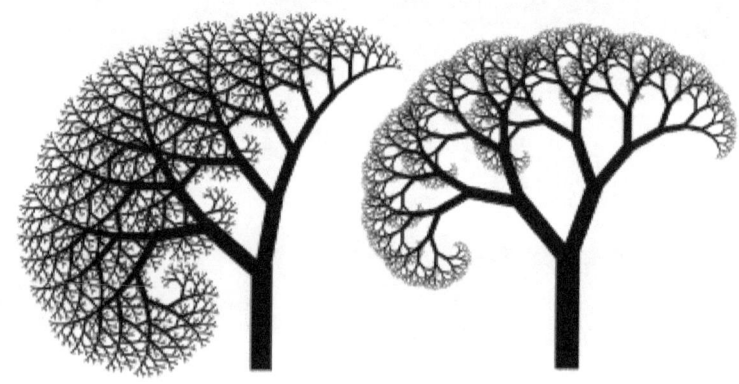

WHERE PATTERNS
BEGIN

PRE-BIRTH PATTERNS are ubiquitous in all organisms
that have sexual origins. Regarding humans, it used
commonly to be believed that a child entered the world

tabula rasa, a blank slate on which his society would write; that is, the child's nurture determined his future.

With our burgeoning understanding of genetics, opinion favored the idea that the child's future lay in his genes. In other words, nature was the architect, and parents did the best they could with the blueprint nature presented to them. Neither argument proved wholly satisfactory, and today the nature versus nurture debate is largely over. We know that both genetics and environment, alone and in combination, contribute to a child's development. And we know that it is not enough to start attending to a child's needs after he is born, because while they are still in the uterus, children begin to lay down patterns.

The first pattern is genetic, and it is strong and immutable. At conception, the father's genetic information is released inside the egg, which contains similar information from the mother. The genes combine; nature's plan is in place. From that moment until the end of the pregnancy, growth progresses at an astonishing pace—especially in the brain and nervous system—and some of that growth clearly involves genetic-environmental interplay and the creation of patterns.

ALTHOUGH THE PROCESSES are complex, involving billions of neurons and billions of linkages among them, development is generally similar for all infants. As the fetus matures, its brain begins to wire itself according to a genetically based pattern. This wiring will not be completed for several years, but the network to support all those neural connections is established in the womb.

Within days of conception, usually before a woman suspects that she is pregnant, the neural tube—the embryonic nervous system—starts to form. The numbers are staggering: neurons are created at the rate of 250,000 per minute until the twentieth week of pregnancy, and by the sixth month the fetus has been equipped with almost all the neurons it will ever need.

Meanwhile, its mother is doing all she can to ensure her child's healthy development. (Optimally, that is. I am assuming she is a responsible woman following her doctor's guidance.)

She neither drinks nor smokes. She may give up caffeine. She "eats for two" in the modern sense, not eating in such volume that she gains too much weight, but being sure that her diet includes all the nutrients she and her child require. She supplements her diet with daily prenatal vitamins and minerals. In fact, if this is a planned pregnancy, she may have begun this regimen even before she became pregnant, so as to give the fetus the best possible start from the instant of its conception.

During the first trimester, the embryo shows crude reflexes and small, random movements. There are the beginnings of autonomic patterning; it swallows and hiccups. Ultrasound scanning detects its heartbeat. Soon, between eighteen and twenty weeks, ultrasound will reveal whether it is a girl or boy. In the second trimester, the child begins to develop reflexive patterns.

Rhythmic contractions of the diaphragm and chest muscles ("practice breathing") that prepare it for respiration after birth are visible, as well as sucking and swallowing—all controlled by the brainstem, which is also the center for vital functions such as blood pressure and heart rate. And early in the second trimester, the mother will feel her baby move. What an extraordinary, life-affirming moment that must be.

Between two and four months' gestation, a great migration occurs: the neurons migrate in groups from the neural tube to locations in the brain where they will take up their various responsibilities. Science knows where they go and what their purpose is, but how is it that these billions of tiny cells know? And with such precision. In fearful symmetry and as though endowed with volition, they move toward their destinations. Their migration is so crucial to brain development that, as I'll discuss later, any disruption of the migratory patterns can have serious consequences for the child before and after birth.

The first neuronal migration is to the interior brain. Later waves of emigrants travel farther outward, and this migration is long, difficult, and fraught with perils. Because other colonies of neurons already have developed entrenched positions in the innermost portions of the brain, those destined for the cerebral cortex—the outer layer and the seat of cognition, memory, emotion, and voluntary action—must struggle through and past them.

For the billions of still-migrating neurons, it's like snaking through a rush-hour traffic jam that extends from Seattle to San Diego. It is just astonishing to contemplate that most of the time these neurons end up where they're supposed to, and brain development proceeds according to plan.

In the last trimester of pregnancy, as the cerebral cortex begins working, the fetus demonstrates its ability to learn. For example, its startle reflex may diminish if the stimulus that triggers the reflex is repeated several times. The fetus gets used to the stimulus and modulates its response. A pattern is being established. The fetus reacts to other stimuli as well. Sensory inputs are rudimentary but clearly present. The fetus explores its environment by touch, moving many times an hour and using its hands and feet to push off from—and walk around—the uterine wall.

It is a weightless world for a tiny astronaut.

What a shock awaits him at birth, suddenly assaulted by bright lights, loud noises, pungent odors, unpleasant touches—all this, and he must deal with the force of gravity too.

No wonder they scream.

But it isn't all completely new. In the womb, the baby was already acquainted with smell and taste. Amniotic fluid carries the odors of food the mother eats, and because the taste buds develop early and the fetus swallows a lot of amniotic fluid toward the end of pregnancy, it probably tastes what she eats too. Some scientists believe that swallowed amniotic fluid is a "flavor bridge" to breast milk, which will contain many of the same tastes; the baby thus arrives more strongly patterned to accept its mother's breast.

Fetuses also respond to sounds and appear to recognize these stimuli after birth. Several years ago in China, Barbara Kisilevsky, Professor of Nursing at Queens College, Ontario, conducted a study of sixty fetuses between thirty-eight and forty weeks' gestation. Kisilevsky, plus a team of psychologists and Chinese obstetricians, exposed thirty of the fetuses to a two-minute tape recording of their mothers' voices reading a poem. The other thirty heard the same poem, except that the voice was a stranger's.

The fetuses who heard their own mothers' voices responded within twenty seconds with an increased heart rate that lasted throughout the recording and stayed elevated for another two minutes. Those who heard a stranger's voice showed a decreased heart rate within twenty seconds, and it remained lower throughout the reading and beyond. These results indicate that a child in the womb forms patterns of recognition: it can distinguish its mother's voice, thereby providing evidence of attention, memory, and learning. The empirical evidence from Kisilevsky's study effectively demolished the *tabula rasa* theory, if indeed any vestiges of it were left.

A variety of other strategies have documented the presence of fetal memory, and there is considerable interest in its purpose. Some believe it to be a practice behavior, like prenatal "breathing." Although most practice behaviors are physiological, if practice before birth helps assure full functioning when the child is born, it isn't much of a stretch to suppose that psychological patterns can be practiced during gestation as well. Research suggests that memory is related to attachment. Kisilevsky argues that because the fetus knows its mother's voice, it is drawn to this familiar stimulus after birth, forming a basis for attachment.

I'm happy to know of the increasing amount of research data about in-the-womb learning; this information adds much to our ability to nurture healthy children. But doesn't it, really, simply confirm much of what already should be obvious?

Have you ever witnessed a newborn lamb, immediately after its birth, struggling on its knees to get to its mother's milk? Or a newborn foal, or a litter of puppies? Having just completed five, or eleven, or two months' pre-schooling in their mothers' wombs, these baby animals seek the warmth and smells and tastes and sounds they are familiar with. They seek the source, the pattern for their prenatal learning: their mothers. And despite our brilliant empiricism, despite our clever tools for examining and reporting upon each infinitesimal step in the conception-to-birth process, it is still a miracle.

EARLY THREATS TO OPTIMAL PATTERNING

IF PATTERNING IS the grooving of stimulus-response exchanges that maintain homeostasis and promote survival, then it clearly begins before the baby leaves the busy world of the womb. The brain is critical to achieving optimal patterning, but because it is so complex and is built over such a long period of time, the developing brain is vulnerable to a variety of external and internal threats.

Earlier, I mentioned that disruption of neuronal migratory patterns can cause serious, permanent harm to a fetus. It is known, for example, that a woman who drinks alcohol in excess during pregnancy may bear a child with Fetal Alcohol Syndrome (FAS), a condition that includes numerous symptoms of a profoundly damaged central nervous system, such as craniofacial abnormalities, developmental delays, and severe mental retardation. These conditions are permanent.

Because it is not known precisely how much and at what point during a pregnancy alcohol exposure causes FAS, the pregnant woman's only safe course is not to drink at all. Of course, alcohol is by no means the only peril to fetal development. Smoking can lead to low birth-weight babies with respiratory and cardiovascular problems. Many other chemical substances pose danger. Babies of drug-addicted mothers, for instance, have themselves become passively addicted in the womb; after they are born they must suffer through withdrawal, a cruel beginning.

These exogenous factors are especially dangerous to the fetus in its first trimester, when neuron migration is at its height. If the migratory patterns are interrupted or displaced by drug use, radiation, poor nutrition, or genetic mutation, the migratory process may be incomplete and neurons "mislaid." For example, some people who received radiation in Hiroshima and Nagasaki while still in the womb were later found to have brain abnormalities traced to incomplete migration. And brain imaging often shows that patients with childhood epilepsy also have pockets of misplaced neurons. If you are in that traffic jam from Seattle to San Diego, and your travelling companion has nagged you into trying a different route, you could wind up like those neurons: at a dead end, or even lost for good.

In addition to everything that menaces the fetus from outside the womb, today we know that one of the most potent threats to its well-being is maternal stress, particularly in the first trimester. Not only is a stressed mother less likely to conceive in the first place, she may be more likely to miscarry. Perinatal psychobiologist Professor of Nursing Vivette Glover, of Imperial College London, notes that a stressed pregnant woman produces high levels of the stress hormone cortisol, which crosses the placental barrier and raises the fetus's cortisol level. A too-high level of cortisol in the fetus can cause permanent central nervous system damage, as "the brain is sensitive to the hormones that are around it, just as it is to alcohol, smoking, or other drugs." Pathik Wadhwa, M.D., Ph.D., Director of the University of California Irvine's Behavioral Perinatology/Development, Health, and Disease Research Program, further corroborates the ill effects of maternal stress: "At each stage of development, the organism uses cues from its environment to decide how best to construct itself within the parameters of its genes," Wadhwa says. "[If] the fetus builds itself permanently to deal with a high-stress environment…, it may be at greater risk for a whole bunch of stress-related pathologies."

Often the media report scientific studies, such as those by Glover and Wadhwa, and almost invariably the reports are oversimplifications. Then, at a second or third or even farther remove from the original sources, readers, TV viewers, and Web surfers extrapolate their own, further simplified versions of the studies' findings.

So some parents, believing that they are giving their child a cognitive head start, offer prenatal stimulation such as poking the fetus at regular intervals, speaking to it through a tube, exposing it to classical music, or flashing a bright light through the mother's abdomen. Most scientists believe that such stimulation does little or nothing to enhance the baby's development, and that in some cases it actually may be detrimental. Shining a light into the womb or introducing too much sound too early, for example, could disrupt the child's sleep/wake patterns as well as its developing auditory system. But although it is probably fanciful to think that singing softly to the fetus will improve its mind, there's a good chance that a lullaby will help its mother relax.

Once acquainted with all this information, some of which can be alarming, what's a mother-to-be to do? She can't live in a bubble. She may have a job, and other children to care for. Probably the most useful advice is the simplest: she should try to nurture herself, and in so doing she will nurture her developing child. Her doctor (whom she hopefully will see regularly) is an excellent source of suggestions regarding her physical and emotional health, and one hopes that her partner is supportive and helpful.

A good diet, exercise, rest, and insofar as possible a stress-free environment—these are the watchwords. There can be no guarantee of a perfect baby, but even under less than ideal prenatal conditions there is still a very strong probability of such. I'm not a gambling man, but given a responsible woman equipped with a skilled doctor, accurate information, and a desire for motherhood, I'd bet on it.

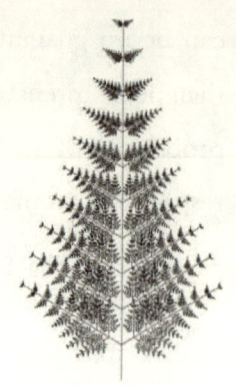

PATTERNING INFANTS
AND TODDLERS

*"Having a child is surely the most beautifully irrational act that two
people in love can commit."*
- Bill Cosby

WHO IS THAT troublesome person who has moved into
your home? He speaks no language that you can understand.
He has no manners and no teeth. His habits of personal
hygiene are deplorable. He is imperious, demanding, and
noisy, and he won't be silenced until he gets his way. You try
your best to attend to his every whim, bringing him food to

assuage his hunger, soft blankets if he is cold, interesting objects to entertain him when he is bored, even your own loving arms to enfold him when he is frightened or in pain. And is he grateful at last? Certainly not. When, after a day of endless catering to this impossible person, you finally fall into bed for some well-earned rest, the tyrant wakes you at ungodly hours, wailing for more attention. Exhausted, you harbor dark thoughts of returning him to that damn stork. But then he favors you with a smile, and, well, you simply melt.

THE BABY'S BRAIN

AT BIRTH, an infant's cerebral cortex is still immature, although the spinal cord and brain stem—the lower brain regions necessary for the essential functions of life—are much further along. These patterns of growth mean that the baby's first responses, such as the startle reflex, are simple unconditioned reactions. Very little cognition is happening. But that is about to change.

The newborn's brain is developing rapidly, building connections called synapses among its hundred-billion brain cells. Driven by a tidal wave of sensory activity, this neural activity begins to refine the brain's architecture. An explosion of synapse-building, known as the "exuberant period," peaks between the ages of six and twelve months.

During this period, as many as two million new synapses may be constructed every second. As the brain awakens, it wires and rewires itself, modifying its genetic expression, strengthening connections crucial to survival, pruning away others, and patterning the growing child for life outside his mother.

It is hardly surprising, then, that during this period a baby may distress his mother with spells of crying for no reason that she can fathom, or surprise her with his delight in something equally mysterious. Misery can switch to joy in an instant without, seemingly, any observable cause. A friend of mine remembers an afternoon when her six-month-old son was sitting on the floor, holding his little blanket and crying. She noticed that the blanket needed laundering, and bent to take it from him, offering a favorite toy in exchange, but the baby ignored the toy and held on tightly to one end of the blanket.

Holding the other end and thinking to extract the blanket gently from his grip, my friend began walking in a slow circle around him, whereupon her son erupted into a deep belly-laugh—something she'd never heard from him before—and kept on guffawing (like, she tells me, a codger who has just delivered what he believes is a real thigh-slapper) while she continued to walk around him until they both got bored with holding the blanket. She says she tried many times afterward to reproduce the blanket-and-circling trick to elicit the same response, but it never worked again and she can't figure out why it worked in the first place.

"He was so miserable," she tells me, "and I didn't know why. And then he was so happy, and I didn't know why." I tell her that if her own brain were producing new synapses at the rate of millions per second, she might seem a little schizoid herself.

NATURE MEETS NURTURE

AT CONCEPTION, the baby's parents provide genetic matrices—genes and combinations of genes that cause or prevent certain kinds of actions—that may or may not appear in the child. The presence of a matrix, even a superior matrix, in a particular area is no guarantee of high functioning in that area. Thus "genius sperm banks" often have disappointing results for the women who pay for their services.

True genius is very rare, and sometimes it arises in a child whose parents are themselves quite unremarkable. By the same token, a woman who purchases "genius sperm" may have an unremarkable child. You can't buy genius. With a judicious combination of good genes and good nurture, however, it is possible to improve the odds for having the bright, talented children you dream of.

For example, some children are born with a genetically patterned matrix for understanding music. If one of these children is born into a musical family, let's say the Bach family, it is likely that his genetic inheritance will be expressed. The child will grow up surrounded by music. His parents and siblings make music, and he too will learn to make music—just as naturally as a carpenter's child will learn to build houses. If a child with a similar musical matrix is born into the Smith family, however, and the members of that family don't value or expose him to music, the child's musical talent may wither. But that is not necessarily the case. Even in the absence of family support or understanding, a child blessed with this matrix often will find a way to express it.

Think for example of Louis Armstrong. The grandson of slaves, the son of an absent father, and a mother driven by poverty to prostitution, Armstrong dropped out of school at age eleven, then helped support his family by singing on the streets of New Orleans. After several minor run-ins with the law, at age twelve he was sent to the New Orleans Home for Colored Waifs, where he received some formal instruction in music and his inborn talent began to flower; he learned to play drums, bugle, and trumpet.

On his release from the institution at age fourteen, he began to haunt New Orleans jazz clubs, where he came under the tutelage of several jazz masters. Despite a chaotic early life, Armstrong's musical genius surfaced; he became a preeminent jazz musician of the twentieth century. The genetic matrix demanded expression, and Armstrong was helped along by those who recognized it.

For him, the time and place and people were right: the early years of the century, New Orleans, the cradle of jazz, the best jazz artists in the world as mentors.

THE CHILD COMES HOME

WHEN A BABY is born, the parents' first job is to do all they can to make sure the child survives and grows, and thus continues the patterns that began in the womb. We are made aware of the baby's great significance even before birth. Ultrasound scans tell us whether the fetus is achieving those all-important milestones within the average time range. We can forecast its dimensions. Many times, if the baby is failing to develop optimally, we can diagnose why and then treat the problem in utero. In fact, what with all the attention that modern science encourages us to lavish on the fetus, we may feel so familiar with the baby before it is born that we expect caring for it after birth to be a fairly simple matter.

But then the baby makes its entrance.

Preconceptions quickly fade, and some big surprises await us. Chief among these is sheer enchantment. The baby enthralls its parents, particularly its mother. This is instinctive. In a University of Wisconsin study, brain scans have shown that a photograph of her infant literally "lights up" a mother's life, or at least her orbitofrontal cortex, the part of the brain that decodes the emotional value of a stimulus. The next big surprise is the alarming knowledge that we have created a hostage to fate, and fate is often unkind.

We must do everything in our power to protect this child from fate's cruelties, and that sometimes, despite our sincerest efforts, we can't. It is then that, humbled, we begin to doubt our ability to rear a hamster, much less a human child. This state of mind is healthy. It makes us more careful, more attentive parents. It helps us get down to the hard, reality-based business of child-rearing.

The birthing process, followed by the baby's actual presence, helps to create bonding and engagement of mother and father. They respond to the baby's "enchantment" stimulus with food, cuddling, and the strengthening of attachment bonds. Such attachment is crucial to the child's survival. Unable to defend itself, it must have advocates. Its strongest advocates are its parents, whose emotions have been so profoundly engaged that they will fight to the death to ensure the safety of their offspring. As its needs for food and attachment are consistently met, the infant begins to trust its parents and by extension its world. And as the child grows older, these healthy templates make it easier for him to function adequately in his society.

But unfortunately not all parents enjoy loving relationships with each other. If the baby's first experiences with his parents are fraught with stress and depression, there can be damaging consequences to his brain and nervous system. For example, researchers at Vanderbilt University and the University of Colorado discovered that if a mother is severely depressed, her speech patterns change: her voice becomes flat and less accented, there are fewer pitch changes and less "baby talk."

It appears also that babies need the alterations in pitch to increase their arousal and allow them to process information more efficiently. Without hearing changes in pitch, babies' developmental patterning is impaired, with a concomitant delay in associative learning and language acquisition. As the widely respected pediatrician T. Berry Brazelton writes, "A threatened nervous system is likely to be hypersensitive and easily disorganized by incoming stimuli

102

and by the baby's own efforts to respond." And a hypersensitive, disorganized nervous system is an adverse environment for efficient patterning.

The baby receives information about how to function, perform, perceive, and feel by interacting with his parents, especially his mother, and mothering begins in the warm, nurturing, protective womb. Breastfeeding continues the attachment of mother and child, and for a very long time as the child grows older, it is still his mother to whom he looks for protection, guidance, and approval.

This long period of human nurturing is rare among mammals. James Kimmel is a former associate of the Natural Child Project, a group that advocates attachment theory in child-rearing. Kimmel writes, "The prolonged mother-child bond [was] the root of human sociability, and the nurturing response of the mother became a model for human interaction. ... We could not have survived as the species we are without attachment to each other." The mother's influence is primary and critical in building these essential patterns of human attachment.

To those of us who have children, or know children, or because we have all been children, this is not exactly headline-making. But if we insist on more evidence, there are psychologist Harry Harlow's rhesus monkey experiments.

In a long series of experiments at the University of Wisconsin during the 1950s, Professor Harlow removed baby rhesus monkeys from their mothers and presented them with a choice of two surrogate mothers, one made of terrycloth, the other made of wire. The baby monkeys invariably chose the cloth mother to cling to. When the wire mother was equipped with a baby bottle of milk, the monkeys went to it for food, but always returned to the cloth mother for comfort. When they were frightened, they clung to the cloth mother for protection.

Harlow concluded that the warmth and tactility of intimate body contact superseded even the appeasement of hunger, at least for rhesus babies. He extended his findings to include human babies, as we and the rhesus are members of the same biological order of primates.

And who can forget the news reports years ago of those foul Romanian orphanages, their young inmates confined to filthy cribs in dark rooms, never knowing a mother's care, receiving nothing by way of nurture from their keepers? We learned of these orphanages because some

Americans had adopted children from them, in some instances disastrously. These tragically deprived children came to their new American homes and their new, loving parents with a host of serious problems: language delays, malnutrition, parasites, tuberculosis, and many more. Presented with toys, they didn't know how to play; they'd never had toys before.

They were dumbfounded by hugs and kisses, and they flinched from human touch. But their most intractable problem, one that some never overcame, was what is known as reactive attachment disorder, or failure to bond. Their adoptive parents' earnest efforts to love and nurture them would never be enough to erase the trauma of these children's previous lives. Their vacant eyes told the story.

I hope that in my zeal to stress the importance of the mother-child bond, I don't seem to be giving short shrift to the father's role.

Not at all.

Although human beings are paired animals, the true human condition is more than one man and one woman. It is a whole system of men and women working within their environment to create their own replacements. As former

First Lady Hillary Clinton has said, it takes a village to raise a child. I believe, though, that a core of at least two persons is optimal for the creation of a high-quality replacement.

The father or partner's job is to help the mother bond with their baby and to defend her against any threats to the bonding process, such as an interfering mother-in-law who may undermine her confidence in her ability to care for the baby. Therefore, during this period of bonding, the partner must provide the necessities of food, shelter, and protection.

Even though the partner may be nervous and confused, consistent, loving support is needed when the mother is tired or frightened or anxious. In other words, the mother must be nurtured so that she can nurture their child. Together, the couple reaches for maturity as they attempt to manage the uncertainty and new challenges of rearing their baby.

In the 1950s, Harlow said that men can nurture their children equally as well as women. At the time, it was a new and surprising idea. I am not certain that the idea is precisely accurate; although nowadays you may see some young fathers sharing in every aspect of baby care except breastfeeding, it remains relatively unusual. Nor am I entirely comfortable with "attachment parenting," at least not when it is taken to an extreme. I've observed young mothers tired out from "babywearing," that is, carrying the baby in a sling around their bodies all day in an effort to ensure bonding, then taking the baby to bed with them at night.

This is an infant care method accepted among many tribal cultures throughout the world, and its salutary effects on bonding are well known.

But most American mothers aren't tribal; they probably had other interests before their child's birth and they will have them afterward. I would caution them to think hard before adopting a method that guarantees two or more years of twenty-four- hour intensive, single-focus child care—something that may not come naturally or easily to them in our culture, that is likely to be exhausting, and that may annoy other parents and alienate their own husbands.

Napping in his own crib, and sleeping there at night, seem to me advantageous for both the baby and his mother. Even in this closest, most loving relationship, people need a break from one another. Perhaps because I'm a product of my own, much older generation, I still prefer to envision the best environment for the child as a strong equilateral triangle with parents forming the bases and the children at the apex. Love and support flow between the parents, creating a solid foundation for a happy, confident child. A mother and father's nurturing roles, in my view, are in many ways separate, but they are distinctly equal in importance.

STYLES OF PARENTING, STYLES OF PATTERNING

THE GOAL OF parenting should be to rear children who are happy, kind, and confident; who feel loved and valued; who are able to enjoy affirming relationships with others; who can earn a living and contribute to society; and who are able to take full advantage of their inborn talents and matrices.

As the patterns form, parents instill values and guide their children toward laudable goals. They may, however, have vastly different ways of doing so. Both parents are working from the patterns they learned and followed as children. Frequently they try to avoid some of these patterns that, now that they have become parents, they see as ineffective or perhaps harmful to their own child-rearing.

A young mother, for example, may remember that her mother was too self-effacing, that she almost always acceded without question to anything her husband demanded. The young mother may therefore decide to be a stronger role model for her daughter. A young father recalls his father as a strict, undemonstrative disciplinarian, so he may determine to try to show own son how much he loves him, or he may continue the patterns of his father.

If the new parents are able to discard the noxious patterns, they combine what are left, forming their own family patterns. Yet despite our sincere attempts to rear our children perfectly, it is disappointing and painful how often, particularly when we are under stress, we revert to the old ways that we hoped we had jettisoned forever.

As a baby grows into a toddler, he is always watching for direction and clues as to how to navigate a world that is expanding for him every day. It follows that a parent needs to become the best role model they can be. Sometimes that is easier said than done, for being a role model asks a great deal. For instance, parents may wish to be a paragon of patience and reasonableness, but having a two-year-old dawdle maddeningly over his breakfast, throw a tantrum because he doesn't like the color of his socks, then squirm so much that you can barely buckle him into his car seat, causing you to run late to an appointment with his pediatrician, your patience may wear thin.

Some parents yell and some use colorful language.

When they cool down, they may wonder guiltily whether they have damaged their child's poor little psyche. They have not. No one is a saint, and after all, losing one's temper after a mounting series of frustrations is a pattern of behavior common to many of us. The child, now happily settling down in his car seat, has learned a few things: how many of his mother's buttons he can push before he destroys her composure (not that he won't try it again, and again and again), how annoying behavior can lead to unpleasant consequences, and maybe a little something about the human condition.

Fortunately for your peace of mind, most patterning for very young children is considerably less existential. Babies can be patterned to go to sleep at the same time every night, and to lie still while their diapers are changed. As they grow older, they can be patterned to use a toilet. If the first thing parents do every morning is brush their teeth and help the toddler brush his teeth, in time brushing his teeth first thing in the morning will become a lifelong pattern. It isn't enough for the parents just to remind the child to brush his teeth; it must be a modeled activity that takes place at the same time every day until it becomes an entrenched pattern.

Waking up in the morning is the stimulus, brushing teeth is the response.

If a child grows up in a home full of books and his parents are enthusiastic readers who read to him, he is more likely to be patterned as a reader. If his parents love music and fill their house with it, the child will be patterned to love music. This is not to insist, of course, that flowers can't grow in a desert.

Louis Armstrong's extraordinary rise out from street urchin to world-renowned jazz artist is but one example. If a baby whose genetic endowment creates an intense desire for knowledge is born into an ill-educated family without an interest in learning—well, not every such child will become an Abraham Lincoln, who had no more than twelve months of regular schooling, but in many cases the child's inborn nature will out, despite his circumstances.

What is of paramount importance in a child's early years is a strong, warm, responsive family. A powerful family unit, parents and children, all working together—doing, sharing, and sometimes sacrificing self to the needs of others —creates the community of civilized adults needed for a warm, responsive, family and society. Bonding with our children is imperative in maintaining a strong society.

As we bond, we pass along not just our own family patterns and values, but also those of our wider society, and we ready our children to take their place in it. And insofar as we serve as templates for our children, we prepare them to be templates for the next generation. It is the only immortality most of us reasonably can aspire to, and it may be the noblest.

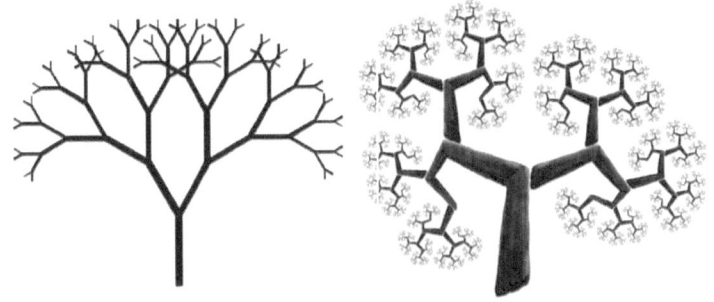

FAIR WARNINGS:
SOME IMPEDIMENTS TO
BONDING AND PATTERNING

AFFLUENCE AND PATTERN VIRUSES

OURS IS A complex society that erects many obstacles to a strong parent-child bond. Perhaps surprisingly, affluence can be one such obstacle.

Always there were exceptions, but in earlier times, gender-specific roles and responsibilities were more clearly defined. Generally speaking, women did "women's work," which by the eighteenth and nineteenth centuries meant birthing and rearing the children and keeping the home. Men were responsible for shelter, safety, and earning. Before that time, in agrarian societies, men and women often worked together in the fields as equal partners, so that women's work was less closely identified only with the home.

In these cultures, children's education was a community responsibility. Parents, religious and community leaders, and schools inculcated the values, beliefs, and mores that ensured the culture's continuance. Often religion was key to enforcement of the group's patterns, but an even stronger dynamic was the society's commitment to survival. Individual rights, therefore, were somewhat limited.

These societies developed their patterns by establishing and compelling compliance with codes of belief and practice. Those who didn't comply became outcasts because, faced with the demanding needs for food, shelter, and protection, the group had little time or energy to waste in dealing with dissent. Today we still can see these societal patterns in a few marginal religious groups.

With the passage of time, as the group's patterns became deeply etched in the minds of its members, they became more efficient. The daily struggle simply to survive began to abate as leisure, and the money with which to enjoy it, increased. As increasing numbers of people worked their way out of hardscrabble lives, they could turn some of their attention to more pleasant things.

Chaucer's Canterbury pilgrims, for example, weren't merely a group of earnest English citizens on a spiritual quest. By no means. Members of a small new middle class, equipped with a little new English money and the desire to spend it, they were also on a holiday junket. Set in fourteenth-century England, the *Canterbury Tales* reflected the England of that time.

Today we see similar attitudes toward life in developing nations, where populations are slowly emerging from direst poverty into comparative affluence. (Affluence is relative. An *Economist* article cites an African Development Bank report that a third of Africans are now "middle-class, defined as having between $2 and $20 to spend a day," and "two thirds of that supposed new middle class have just $2 to $4 per day.")

So what's the matter with affluence?

Most people wish they had more of it. Well, I have no wish to impede progress, but I do think we should be more alert to its perils. One of them is what I call a "pattern virus," that is, a deviation from the accepted patterns of the group; it is a stimulus based in experience and "injected" into the child through modeled behaviors of parents. A biological virus reproduces by injecting its genetic material into a living cell, reprogramming the cell so that it reproduces the virus instead of itself.

The common analogy is to a computer virus—and you probably know only too well how much havoc a virus can wreak on your computer, and indeed on a massive, nationwide network of computers. Another analogy is this: People are to a society as genes are to the body. As the "genes" of an established social order, members of a society pass its patterns on from generation to generation. And just as a biological virus can infect the body, a pattern virus can infect the body politic.

Racial and religious prejudices are among the easiest pattern viruses to understand. Parental experience is often shaped by encounters with people of different racial or religious backgrounds, and they inject (pass on) their belief regarding those people into their children. If enough members of the community come to share the belief, more of them pass it on to their offspring. A social pattern virus flourishes, to be transmitted generationally like a genetic trait. Transmission requires that a society's elders pass the virus on to its children, who are born free from prejudice.

As the song from *South Pacific* explains so succinctly:

"You've got to be taught to be afraid / Of people whose eyes are oddly made / And people whose skin is a different shade." And the sooner the better: "You've got to be taught before it's too late, / Before you are six, or seven, or eight, / To hate all the people your relatives hate."

Yet for all its ability to infect many successive generations, a pattern virus is fragile. If only one generation does not express it, it dies out. You would think, then, that by this point in the evolution of the most intelligent species on earth, we might have eradicated all the viral bigotries that have so often been agents of our destruction.

You'd think that, just for a generation, we might all shut up long enough to free our children of preconceptions and biases, and let them to get to know and like each other regardless of race, religion, or any other difference. No such luck. Millions of years after our emergence as *homo sapiens*, we remain infected. It is pandemic. And it appears to be a permanent part of the human condition.

The connection between affluence and pattern viruses is this: affluence has nearly always resulted in new pattern viruses. As I've said, with less need to work so long and hard, members of affluent societies had more time to acquire property and greater wealth, and to consider the advantages of individual rights.

Over a few generations, the pattern of communal rights eroded, to be replaced by personal rights. In our society, that new pattern continues to hold sway; it affects—not always negatively—every important aspect of our culture, and its most deleterious effect is on mother-child bonding.

Among the pattern viruses related to mother-child bonding that arose, particularly in Western civilizations, was the belief that the rights of the mother superseded those of the child. Indeed, it was thought that mothering itself was dispensable. Wet nursing of infants is an example: an accepted practice since ancient times, most often used when a mother was for some reason unable to lactate. It was customary among aristocratic or other very wealthy families.

In Europe, particularly during the seventeenth and eighteenth centuries among prosperous families, and among families who wished to emulate them, wet nursing was a very common method of child care. Mothers farmed out their babies to wet nurses, often for as long as four or more years. The popularity of wet nursing spread to people lower on the social ladder, as the ability to pay for a wet nurse became a sign of "class."

One result of wet nursing was the startlingly high birth rate among women who made use of a wet nurse's services. Lactation suppresses ovulation among most women, but those who don't breastfeed can become pregnant again within a month or so of giving birth.

And they did.

Bearing seventeen or eighteen children within twenty years was not unusual. Twenty years of nearly continual pregnancy were certainly much more debilitating, more fraught with health risks, than breastfeeding would have been. For these women, "freedom" from those pesky babies—which included foregoing the mother-child bond augmented by breastfeeding—came at a high price.

Freedom from the demands of nursing and infant care was perhaps liberating for some women, but an unintended consequence was that sometimes wet nursing severed the attachment patterns that had formed in the womb or immediately after the child's birth. Some mothers lost that intimate connection with her baby, and the baby, deprived of her nurturing, was set adrift.

The wet nurse herself was doing the job of feeding the infant for wages and seldom felt emotionally invested in the child. In fact, she might be nursing several children from several different mothers simultaneously. Bonding, if it occurred at all, was likely to be superficial and related only to nourishment—rather like the baby rhesus monkeys' cursory attachment to their wire surrogate mothers.

Moreover, the mortality rate among these children was quite high. They died for any number of reasons, a particularly lethal one of which we now call "failure to thrive." This term includes retarded growth, delayed intellectual and physical skills, diminished human interaction—a seeming lack of life-force. Failure to thrive can occur in any socioeconomic group, but we see it most often among neglected or abused children, institutionalized children, and children living in extreme poverty. Often, doctors are unable to pinpoint its causes.

Maybe, sometimes, it is simply heartbreak.

Small wonder that without a pattern of bonding with its mother or wet-nurse, a child finds it difficult to form attachments with anyone else. The patterns of nurturing that mothers traditionally provided were broken. The development of empathy, tenderness, caring, and compassion in the child had been sacrificed to individual rights and personal freedom of the mother.

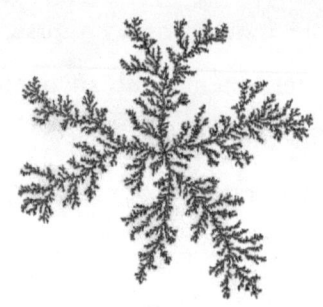

A STORY WITH
NO END IN SIGHT

IN SOME WAYS, this story is our narrative as well. As American society has become more efficient, we've also become more affluent. With more leisure, our focus has shifted inward, away from the values of the family and the group. Individual rights and freedoms have assumed greater importance, and with this change in emphasis, the tight bonding of parents and children—especially between mothers and children—has begun to decline. American public policy, unfortunately, supports this pattern. Of all the

122

developed countries, the United States provides only brief family leave, forcing working mothers and fathers away from their newborns within weeks of their birth.

The Family and Medical Leave Act of 1993 provides only twelve weeks of job-protected, unpaid leave for those who work for firms with more than fifty employees and who have worked at least twelve months for that employer in the preceding year.

Much vaunted as a major step forward in America's societal welfare progress, the Act fails to cover a huge proportion of the workforce, and as it provides no income replacement, many who are technically eligible don't take the leave because they can't afford to do so. Admittedly, the Act was a small step in the right direction; at least some people no longer faced the spectre of being fired for becoming parents.

The United States is one of only four developed countries that does not have a national law mandating paid time off for new parents, although the Family and Medical Leave Act of 1993 mandates 12 weeks of job-protected leave (potentially unpaid). Also, individual states can mandate paid family leave, including parental leave for same-sex partners. In some nations, mothers may take three years to bond with their children before returning to their jobs and generous family leave also is available to fathers. These policies promote a longer period of attachment and an atmosphere more amenable to laying down patterns of nurturing. Safe in the knowledge that they are financially secure and that, when the time comes, the job they left, or a similarly remunerative one, will be there for them, new parents can relax and take care of their babies. And it must be a boost to their self-respect to know that their employers and even their governments think they are doing a very important job at home.

Beyond public policy, parental patterns also contribute to a breakdown of family structure. Parents may be working in professions that take them, in physical and/or psychological distance, farther and farther from daily interaction with their children. Parents who don't work may prefer to pursue their own interests rather than stay at home. In both cases, child care is turned over to caregivers or daycare workers.

For single parents, there is seldom a choice. They must work, possibly at two or three jobs, just to put bread on the table. Many parents, however, do have a choice, and too

124

often they choose their own self-development over the needs of their children. Some daycare workers to whom we entrust children may be warm and loving people, but they aren't the parents. Abused and neglected children sometimes can find salvation in the arms of a caring, concerned daycare provider, but as a rule the daycare setting cannot fulfill a child's attachment needs.

I don't mean to suggest that daycare is no more than an advanced form of wet nursing. Most daycare facilities, at the very least, follow basic rules of sanitation and worker-child ratios. Some are models of good child care, with employees who have degrees in early childhood development, low worker-child ratios, spacious quarters well equipped with books and instructional toys, as well as plain, old-fashioned crayons and blocks. Other facilities may not be so generously appointed, but they are staffed with people who, while they may lack college degrees, make up for it with genuine affection for and understanding of young children.

Yet, in too many cases, some of the parallels with the old practice of wet nursing are disturbing. Sometimes what passes for daycare is no more than a neighbor down the street who takes in a dozen or so children just as a means of earning money. The "care" offered amounts to little more than installing children in front of a big television set for most of the day. And money itself is an issue; typically, wages for daycare workers are pathetically low, making it hard to attract committed staff, thereby encouraging high employee turnover. This often forces the daycare director to hire unqualified workers.

The neighbor down the street may be only a babysitter, offering minimal supervision and no attachment. Not even an ultra-skilled daycare worker can hold and solace every child who needs comforting. Daycare providers are *paid* to look after our children. Money—especially those pervasively low wages of daycare employees—is a poor incentive for the good care and nurturing of children. The results—the parallels—are clear. Daycare facilities often have too many children to attend to adequately.

Each of the children is contending for love and attention. Each wants to bond, and each may fear rejection. Early development involves patterns of fear and uncertainty. Too frequently, children's ability to understand others may be stunted. They may not learn to empathize, they don't learn to love. A diminished ability to love is the hallmark of a child whose bonding is incomplete.

As poor bonding moves from generation to generation, the society weakens commensurately. People who don't know how to care about one another are less able and less likely to come together for the common good. How many news stories must we hear about whole neighborhoods ignoring a rape victim's cries for help or a victim of a hit-and-run accident lying bleeding in the street, before we realize as a people that something is very wrong here?

Governments that refuse to care about their citizens can't expect them to hold together in times of crisis. How many times can our politicians dupe us before we lose all trust in government? How many times will our young men and women volunteer to face death in wars once they learn the cynical motives for starting them? How long will it be before our Congress takes another, *real* step forward, one that respects the essential right of parents to have time to bond with their child?

I hope that it is too soon to speak of America's decline and fall. I hope that we can do better. Yet all empires fall. Other societies, with stronger, more robust patterns of belief, a healthy work ethic, and genuine commitment to its children's welfare, topple them. Our children are our future, as contemporary society likes to say. But unless their parents nurture them, bond with them, and teach their children essential human values, our future will be bleak.

After the near-collapse of Wall Street, we learned who had caused the disaster: a few men of towering selfishness, greed, and egocentricity. Look at the multi-billion dollar payout by JPMorgan Chase to the government in 2013. Millions of Americans suffered because these supposed guardians of our financial system, which is based on trust, failed in their responsibilities. These men sapped our economic strength, stained our international reputation, and monstrously betrayed our trust. The government rescued some of them, and even if we understood that necessity, we still felt that we were being deceived, and that prison time would have been a more satisfactory option. In other words, the American people lost faith in their institutions, and for most, it has yet to be restored. This augurs ill for our future as a united nation.

I can't say, of course, how those Wall Street magnates were brought up. I don't know what their early childhoods were like. Maybe they *were* taught values, and then discarded them somewhere along the way. It might be illuminating to find out.

PATTERNING CHILDREN AND ADOLESCENTS

Between the dark and the daylight,
When the night is beginning to lower,
Comes a pause in the day's occupation
That is known as the children's hour.
I hear in the chamber above me
The patter of little feet,
The sound of a door that is opened,
And voices soft and sweet.
From my study I see in the lamplight,
Descending the broad hall stair,
Grave Alice and laughing Allegra,
And Edith with golden hair.
- "The Children's Hour" (1860)

HMM. WELL, LONGFELLOW'S vision of idyllic family life may have been accurate for him. After all, to a very famous and very prosperous New England poet, 1860 was a kinder and gentler time than our own. There were servants and a wife to care for the children throughout the day so that, undisturbed in his study, Longfellow could attend to his remunerative poetry and speaking engagements and European tours and various professorships.

His "children's hour" was probably just that: "a pause in the day's occupation" to relax and enjoy the company of his three daughters, bathed and nightgowned and ready for bed. In Longfellow's comfortable circumstances, fatherhood appeared to be all roses.

But alas, for most families today, life is quite different. The children's hour lasts either all day long, or it is crammed into the time before and after school or daycare, and it is seldom idyllic. Say, for example, it is after dinner at a typical household. Thirteen-year-old pouting Alice is refusing to speak to her parents because they won't allow her to hang out with friends at the mall unsupervised. Nine-year-old sobbing Allegra has shut herself in her room because her mother won't buy her a cell phone. Four-year-old wailing Edith is devastated because her father made her stop "helping" him in the garage while he was using power tools. What would Longfellow do? Probably bar the door to his study.

PATTERNS OF PLAY

ON THE SPORTS field there may be two mixed teams of children, boys and girls, six- and seven-year-olds, playing soccer. Some are in red (the Sheldon Lighting team) and some wear blue (the Samson Movers team). Their coaches are adults—most likely parents. It's a classic Little League scene and a common pattern for forming team alliances.

Most children feel comfortable associating with their peers in an identified group. These playground soccer kids are more than soccer kids; they are teams. Something as simple and symbolic as similarly-colored T-shirts sporting the team's name can give these children an identity and create a cohesive group with a strong sense of commitment and importance. Take a group of prepubescents, dress them in uniforms and teach them a game's basics for three weeks, and they will do almost anything for their team. These children are developing patterns that they may apply in adulthood.

But some children don't see the point, or the fun, of chasing a ball around a field. They would far prefer to spend their time drawing pictures or writing stories or taking solitary, thoughtful walks. They aren't antisocial; they are likely to have one or two close friends. These children too are forming patterns that carry over into their adult lives. Because they are not naturally "team players," they require a different kind of "coaching"—the kind that encourages their individual gifts rather than developing group behaviors. I sometimes wonder how many budding Michelangelos or Beethovens or Nureyevs have been stymied before they even reached their teens by parents who forced them onto the sports field.

In any case, change is coming for all these children. Their bodies and their patterns will change. They will enter adolescence.

Humans have much in common with the great apes. In fact, chimpanzees are more like humans than they are like other apes: their DNA sequences are 99 percent identical to our own.

However similar we are, though, that other one percent creates enormous differences. One is that human beings have a much longer period of neoteny than the apes. That is, we retain more juvenile characteristics late into adolescence, which delays our maturing.

After birth, the ape's brain continues to develop for a short time and grows about twenty percent. The human brain doubles in size in the first year, and by the time a child is six, his brain weight has tripled. These are the crucial years for developing billions of synapses and forming patterns. Humans develop much larger brains than apes, and there is a clear relationship between brain function and the number of cells and synapses involved.

Stimulation of the child's brain during these early years is important for maximum development. Without knowing much at all about cells and synapses but relying instead on observation and common sense, effective parents have enhanced brain stimulation, providing their child an enriched environment. The child learns to name the parts of his body. He learns shapes and colors and textures and sounds. When the time is ripe, he learns letters and numbers. He learns to read and to count. At six or so, he may begin piano lessons. *Piano lessons!* And in only six years. An unimaginable number of synapses must be in place for this miracle of mind and sensorium to occur, and prodigiously more must develop before he becomes the next Horowitz— or maybe just plays a little ragtime.

A young ape's body matures much more rapidly than a young human's body. Ape babies and human babies are straight-backed, with erect heads. But the ape's maturation occurs faster; its back bends and its head moves forward relatively quickly; a fully mature body is evident much earlier. From this point the ape's intellectual development goes, comparatively, nowhere, and we leave our cousin in the dust.

In humans, however, the aging process is held back, and this slower maturation and protracted adolescence give humans a long time to lay down complex patterns. Moreover, parents have more opportunity to exert their patterning influence on the child, and so do the groups—family and community—to which the child belongs.

The infant's brain is infinitely creative: Babies are continually creating new thoughts, acquiring new knowledge, and experimenting with new activities. They are voraciously curious, patterned by nature to take in everything in their environment. And they are remarkably non-judgmental, as ready to eat dirt (or worse) as to drink milk. Parents must "baby-proof" the house well before their child starts to crawl, and then double and redouble their efforts to keep him safe as his curiosity and range expand. To a baby, life is one grand, enormous Whitman's Sampler.

Infants also are perfect narcissists—the center of their universe, the sun around which all the planets revolve. Parents and grandparents dance attendance upon these little tyrants, and for a while that is a good thing, because apart from screaming for attention, a baby is helpless to take care of himself. In a couple of years, instead of screaming he may take up whining, or bullying, or some other irksome method of getting his way. (Some people, in fact, never mature past this narcissistic stage; they go on screaming or whining or bullying their way through life. Sometimes this gets them what they think they want, but it's no way to win friends.)

The parents' responsibility is to seek a balance somewhere between overindulgence and overdiscipline—a challenging job, for as the child grows his needs are always changing. And then, at the very time when parents may be congratulating themselves for finally having got this parenting thing straight, adolescence strikes.

The adorable, cooperative ten-year-old turns, almost overnight it seems, into someone you barely recognize, and all bets are off. Perhaps the most practical use of neoteny is to remind parents not to throttle their adolescent offspring. Through clenched teeth parents tell themselves that *he is still a child*.

Early in life the carefully reared child abandons some of his narcissistic ways and adopts the patterns that his culture and society expect of him. Thus demands give way to polite requests and sharing supplants grabbing. Slowly the child learns that immediate gratification is not always possible, that other people's needs are important too. Glimmerings of altruism appear. The time it takes to make these steps toward maturity varies according to the genetic and experiential patterning of each child. A woman I know, for example, has a four-year-old granddaughter whose genetic template is largely unknown, as she was conceived by means of a donated egg and donated sperm. That is, she carries the genes of neither her mother nor her father, but of the anonymous donors.

The grandmother tells me that little Eliza's behavior, unlike her own children's, has been "quite challenging" (read: the kid can be a holy terror). But the grandmother also tells me that Eliza's mother tries to deal with her in ways that enhance her strengths (athleticism, keen intelligence and curiosity) and curb her faults (impatience, tantrums) without diminishing her delightful *joie de vivre*. Eliza attends an excellent pre-school and loves her swimming and dancing classes.

Consistently-applied time-outs have reduced the tantrums. Extending by slow degrees the length of time Eliza must wait before getting what she wants (*right now!*) has begun to teach her patience. These are the sorts of things that loving parents do to help their children become valued members of society. Gradually, and by various means, the rough clay of infancy and early childhood is being shaped into what people generally acknowledge as a civilized human being.

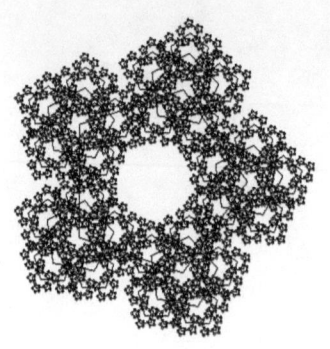

EDUCATION AND PATTERNING

THAT OUR PUBLIC schools seem to be failing our children should come as no surprise to anyone who has been an administrator in them, taught in them, or served a twelve- or thirteen-year state-imposed sentence as a student in them. I know that there are some impressive exceptions—however, I am speaking of the majority of our public schools. The statistics are undependable—estimates of numbers of dropouts per year, for instance, vary according to the agendas of the sources who produce them—but uniformly grim.

Observation and common sense, however, can tell us a lot.

Today's readers of history, for example, marvel at the high level of literacy apparent in first-hand accounts from the Civil War, many of them written by young soldiers who had no more than a few years of elementary education, if indeed they had been to school at all. But a friend of mine who graduated from high school in 1958 and who now teaches English at a university told me, "Mine was a pretty typical American high school, and back then about ninety percent of a typical graduating class were prepared to do college-level work, even though a relatively small number went on to college. Today the numbers are reversed. Many more high school graduates go to college, yet about ninety percent of our entering freshmen have to take remedial courses in math and English."

Once, I strolled through the halls of the engineering and mathematics buildings at a university. Photographs of recent graduating classes hung along the walls; nearly every one of the graduates was Asian. "It's been this way for decades," an engineering professor told me. "They come here for their education, then take it back home to their own countries."

Although parents are usually a child's first teachers and strongest pattern shapers, when the child reaches school age, the school takes over much of the patterning process. School is not the only place that reinforces communal patterns, but it hopefully is the arena where children's patterned matrices are able to flower.

It takes a combination of the individual child's matrix, skilled teaching that instills a pattern for disciplined work, and emotional support, for a schoolchild to develop his full potential. A good school can do much to ensure that these requirements are in place.

More and better testing is needed to identify children's innate, genetic matrices in order to tailor their education in ways that encourage them to achieve their potential, and/or that compensate in those areas for which they have no natural affinity.

Each child is different.

Each child can be reached by different approaches.

It is rather foolish, then, to expect all children to respond well to the same curriculum. "After we got to long division I wept over my arithmetic homework every night, all through school, and I felt really stupid," a writer told me. "Then, in college, I even flunked the course everybody called 'Math for Dummies.' I just knew I'd never graduate, and all because of that one stupid course." Fortunately, an understanding advisor arranged for her to substitute a graduate-level linguistics course (not exactly 'Language for Dummies,' I'd say) for the math requirement, and she excelled in it.

It may turn out eventually that schools will specialize in areas that address students' particular matrices instead of insisting that every student master a broad-based curriculum. Some cities already have schools for students talented in the arts. These schools don't ignore the basics, of course. Even budding thespians and prima ballerinas need to know how to handle their checkbooks. But if the rigors of, say, Geometry 2 discourage a gifted young soprano from applying to Julliard, then maybe her school needs to refashion its curriculum.

It well may be that many students' problems that are perceived as psychological are caused by a lack of understanding of a child's variety of genetically patterned matrices. When Thomas Edison began school, his mind

143

seemed to wander a good deal—so much so that his teacher, a Reverend Engle, labeled him "addled." That comment ended Edison's three months of official education; his mother withdrew him from the school and taught him herself, at home. I'd bet that what caused the young Edison's fertile mind to "wander" was boredom—the sheer tedium of the good Reverend's classroom. What a loss to our world it would have been if Edison had stayed in that school, and boredom had turned him into an unhappy, underachieving, shunned misfit.

Our schools ought to serve as cradles of creativity. Mostly, they don't.

Although parents and schools pay lip service to the idea of creativity, when it actually shows itself it is frequently quashed. Schools want obedient, amenable, well-patterned children who are creative within acceptable limits, rather than children who are burdened with the baggage that often accompanies the strong creative impulse. They want, in other words, children who can take their creativity or leave it alone. But the creatively gifted child *can't* leave it alone, and should not be forced to.

Recall those thoughtful, rather solitary children I mentioned at the beginning of this chapter? The constraints of ordinary classrooms can sometimes damage them, sometimes irrevocably, so that their extraordinary gifts wither. Although these children may seem difficult and hard to reach—often "off in their own world" and unresponsive to classroom requirements—the responsible school doesn't punish them or simply leave them to their own devices; it encourages and guides them toward excellence in the areas in which they show great promise.

And highly creative children can distress parents who perceive creativity as no more than a pleasant aspect of a well-rounded child. Yet if parents truly wish to foster their child's creativity, there are several ways, their efficacy dependent on the needs of the individual child:

- *Give them time to create.* They will learn soon enough that the clock dominates our lives. But creativity often means ignoring the clock; sometimes the child needs to follow their creative process to the end without regard for time.

- *Let them taste and question and sample and experience.* Of course you have to forbid what is dangerous to children, but to a safe extent, and age-appropriately, allow them to skirt the edge. This will help them as an adult, perhaps on the trail of some groundbreaking discovery in their chosen field, to recognize the precipice ahead of them, acknowledge there may be rocks below, and judge for themselves whether the outcome is worth the risk. It is this kind of decision that brilliant attorneys, businesspersons, artists, writers—all people who excel in their disciplines—must make.

- *Encourage them to try.* Fear is normal, but it can block creativity. Children fear rejection, ostracism, loss of their family's love if they fail. Reassure your child that making the effort is admirable, and that failing—maybe failing many times—is just a normal part of trying. Remind them how long they took to learn to ride a bicycle or swim, and what a skilled rider and swimmer they have become, and how much you love them no matter what.

- *Give them discipline, not punishment.* Real discipline is not punishment, although too many schools apparently believe it is. Discipline is teaching; you can teach and model self-control, respect for self and others, courtesy, and all the other attributes that help your child make the most of their gifts. And you can teach intellectual discipline, without which creativity dissipates. (You'd better, as our schools appear incapable of doing so.) Start early to teach them to think critically.

All children have creative potential, but some children's gifts are so powerfully integral to their genetic patterns that, despite the attempts of parents, schools and society to enforce conformity, creativity will fade. Societies seem to recognize the need for creative people, yet do little to encourage and nurture them.

To many conservative members of Congress, the NEA and the NEH are totemic: to the members' minds, they mean social destabilization. This thinking, in fact, is to some extent accurate. The new, exciting ideas that flow from creative people are often subversive of the established order; new ideas disrupt social patterns. A society seems intuitively to know this, and in its efforts to protect the status quo exerts harsh pressure on creative individuals to march to the community drummer. Many fall into step. The result is a winnowing out of all but the strongest, most gifted, most self-motivated children—those who, having survived this Darwinian process, may evolve into truly great artists and writers and inventors and thinkers. Ironically, their work and influence will enrich the very society that made their journey so hard.

A highly creative child may throw sand in the gears of a smoothly functioning school, creating disorder in his need for a different kind of learning experience. Teachers may ignore him, or even punish him for expressing his genetic endowment, and grades are most often their weapon of choice. Learning is supposed to be a cooperative venture for teachers and their students, but too often it turns into an adversarial relationship, with grades giving the teacher the clear advantage. I know a high school math teacher who, regularly and with obvious relish, flunks forty to fifty percent of her students. She seems in fact most content when she is able to flunk a struggling senior who needs to pass her course to graduate. What kind of educator is that?

Highly creative or not, children tend to live up, or down, to expectations. And all children need to feel that the work they do in school is important. We need teachers, therefore, who take their students' work seriously, who judge it carefully according to its own strengths and weaknesses, and who offer constructive praise and suggestions for improvement.

Instead of imposing letter grades that persuade the A student to believe he is a superior person, the C student to think he is an average person, and the F student to know for certain he's a failure, why not do away with grading? Why not, instead, give each student's work the sort of attention that, no matter at what level he is achieving, convinces him it is important?

Rather than assigning letter grades, which mainly reveal whether a child is good or bad at passing tests, why not, for example, write thoughtful evaluations of his work and discuss them with him? A handful of private schools and some colleges use this method with excellent results. And rather than force a young child who has "failed" to repeat a whole academic year, why not, in the first place, allow him to progress at his own pace, moving forward quickly in some areas, remaining longer in others? (No one who failed first or second grade is ever fully free of the feeling that, despite what success he may have achieved throughout the rest of his life, he is really still a "failure.")

Given the demands of top-heavy school administrations, of too-large class sizes, of one-size-fits-all standards, and of too many tests that overworked teachers must teach to, perhaps these ideas seem quixotic. I disagree. I think they are, in the best sense, practical; that is, not only can they work for almost every student, but they also can improve our society. The great barrier to their implementation is finding out how to transform the mindset of that lumbering behemoth known as the American public education establishment.

ADOLESCENCE AND THE BRAIN

THE TRANSITION FROM childhood patterns to adult patterns—the period known as adolescence—is a time of profound change and high risk. The brain is undergoing dramatic changes. It used to be thought that the human brain's astonishing growth and development was a one-shot wonder and that the brain's structure was in place and relatively stable after the age of six.

This belief often caused parents much anxiety, as they assumed that the die was cast at an early age, and that the personality essentially was formed by the age of three. The work of many scientists, probably the best known of whom were Swiss psychologist Jean Piaget and Russian psychologist Lev Vygotsky, has shown conclusively that children continue to develop and change from infancy through adolescence, and sometimes into their twenties. And although, as I've stressed, early neural connections are of critical importance, recent brain imaging studies show that the brain continues to organize itself for a much longer time than previously believed, so that the years between early childhood and late adolescence are also fertile periods of neural development.

This information should come as a relief to parents who fear that the mistakes they made with their small child may have scarred him for life. There is still ample time for them to rectify matters, as well as plenty of neural activity to equip their adolescent for control of his own destiny.

Brain development occurs in a cycle. First comes the production of gray matter—neurons that are not yet part of neural circuitry. These cells form neural networks that are based on experience. These are the bases of patterns, and those patterns are dictated by a sort of "use it or lose it" mechanism. If a child is studying music, for example, his brain will strengthen the "musical" connections that are reinforced as the child practices music daily. After the gray matter is organized there is a period of pruning, during which the brain's white matter—the tissue through which messages pass between different areas of gray matter—reinforces and stabilizes the most vigorous neural pathways, while the weaker ones are pruned away.

From age three to age six, the growth in children's brains occurs in the frontal lobes, but from age six until the beginning of puberty, which usually occurs before the teen years, gray matter increases in the temporal and parietal lobes. Then, with the onset of adolescence, though more development of gray matter occurs in other areas of the brain, there is a loss in the frontal lobes.

Parents agonize that their teenagers are as volatile and emotionally unpredictable as toddlers. There's good reason for this. The frontal lobes govern impulse control and emotional regulation. It is not surprising that these areas grow in early childhood, while children learn to live in society and modulate their emotions and temper tantrums. It is also sensible that as these areas show shrinkage in the teen years, young people's emotions once again become raw, and confusing both to them and to their parents.

Yet in spite of all the *sturm und drang*, adolescence is, as I said, a time of opportunity for teens and young adults to exert some control over the development of their own brains. During this period of pruning and new growth, if teens conceive a particular interest in something—anything from art to zoology—then intensive work will hardwire their abilities in that field.

I recall, for instance, a young man in my high school class. Adam, an indifferent student, drifted aimlessly through his freshman and sophomore years; nothing, apparently, could engage his interest. But in the summer before his junior year, his grandfather persuaded him to join a bird watching expedition. It changed his attitude and his life. Adam began to apply himself to his schoolwork, not because he suddenly found it interesting, but to get the grades he needed for acceptance by the university where he'd be able to follow his bliss. He became—what else?—a renowned ornithologist.

ADOLESCENT
PATTERN CHANGES

IT ISN'T JUST the adolescent's brain that is growing and changing. Autonomic patterns are shifting. Hormones are being released, the body is growing and its shape is changing, bones are hardening, sexual characteristics are becoming visible. And with the new body comes the potential for pattern change.

In order for adolescent and adult patterns to form, childhood responses to stimuli must diminish. The childhood response pattern is always present, even in those of us who seem fully mature, but a purposeful adolescence alters those old patterns, and the teen begins to respond in more adult ways. It's important to know that because the earliest patterns remain in the matrix, regression is always possible. Adolescence is a time of flux; new patterns are not fully set, so that the same stimulus can trigger an adult response or that of a five-year-old.

Because of the capricious character of adolescence, it can be impossible to predict which response it will be. Last night, for instance, you asked your teenage daughter to set the table for dinner. She always had complied willingly with the very same request, last night it provoked an explosion: "I'm treated like a servant in this house! Will you quit nagging me about everything?!" Then your erstwhile reasonable child stomped upstairs to her room and slammed the door, leaving you mystified as to what you had done wrong.

Patterns established in childhood are hard to change and without considerable emotional upheaval will likely remain fixed. And since emotional upheaval defines adolescence, it is a prime period for pattern shift. Adolescence is in fact the only time in life when it is normal for patterns to change. Forged in the heat of adolescent rebellion and emotionalism, the new patterns are strong. Usually, that is a good thing. Our own American history provides an analogy.

People tend to imagine our "forefathers" as graybeards, but in fact most of them were young men, eager and ready to break their old ties with King George III—that is, the Colonies' childhood patterns of obedience and dependence. It was a momentous step, all the more thrilling because these rebels were still British subjects, thus answerable to the laws of treason. We forget too that many American colonists were staid Tories, establishmentarians loathe to rock the royal boat.

Nevertheless, led by youthful men of vision and the courage of their convictions, the rebels prevailed, as youth and vision and courage do. Now Americans could develop their own unique identity. They would be free to pursue their own lives, their own liberty—and of all the newfangled, unsettling patterns issuing out of those tumultuous times, this last one had to be the most startling—their own happiness. I believe that is called growing up.

Adolescence, then, holds great potential for a good outcome, and plenty of risk for a poor one. If an aberrant activity, such as substance abuse, is introduced during this period, the dysfunctional behavior may persist throughout life. Once the pattern is in place, it becomes part of the organism, almost impossible to remove without a tremendous emotional (and, often, financial) investment by the sufferer and his family and friends. In my family, for example, Uncle Joe was the aberration.

When Joe was fifteen or sixteen, he incurred gonorrhea, the treatment for which left him in chronic severe pain. His doctor addressed the pain with laudanum, an opiate. The result was that Joe became an addict. When the federal government placed stringent restrictions on the sale and use of narcotics, Joe became a criminal, and unable to support his habit, Joe also became a gambler.

Despite the family's unending attempts to help him with every resource available at the time, Joe remained narcotics-addicted. His addiction to the drug caused major pattern changes that deviated from the normal adolescent and led to patterns that enabled him to survive as an adult. Joe was always on the run, in and out of prison, denying or exploiting his criminal connections, forever chasing the next big score. He was a brilliant, charming guy. "Joe could be anything he wanted to be," everybody in the family said, "if only...." They didn't have to finish the sentence; we all knew it was hopeless. His survival patterns served to keep him alive until he was in his forties, when he died of tuberculosis.

PUBERTY RITES
AND ADOLESCENCE

היום אני אדם

"TODAY I AM a man." So begins the thirteen-year-old Jewish boy's ritual speech marking his passage from childhood into adulthood.

160

Hozho nahasthlii. Hozho nahasthlii. Hozho nahastlii. Hozho nahasthlii.

Walk in beauty. Walk in beauty. Walk in beauty. Walk in beauty.

So ends the benedictory prayer for a young Navajo girl who is entering womanhood.

The boy's Bar Mitzvah is preceded by several years of studying Hebrew, learning the Torah and the Talmud, and practicing the ways of the religion, and followed with great celebration and gifts from his family and friends. The Navajo girl's *kinaalda* consists of four grueling days in which she fasts, prays, runs a mile at dawn in all four directions every morning, demonstrates her skill at traditional tasks such as grinding corn, carding wool, and butchering a sheep, and engages in a variety of other demanding activities and rituals. The ceremony is accompanied by great celebration among her family and friends, who bring gifts for the celebrant.

Because people recognize adolescence as such a crucial period in human development, many cultures have established rituals with which to move its youth from their adolescent patterns into healthy, culturally productive, mature patterns.

The ritual forms differ, depending on what characteristics the culture wishes its men and women to embody. A hunting culture, for instance, seeks men of strength and endurance; a warrior society looks, in addition, for courage and selflessness. All cultures seek potency in their men and fertility in their women. Some kind of ordeal is commonly a component of the rituals, and some ordeals are brutal, involving genital mutilation, piercing of body parts, stringent fasting, prolonged seclusion, beatings, tearing of one's flesh to appease the gods.

Indeed, just about any torture—a few so multiple-X-rated that I forbear to mention them and try never to think about them—that the human mind can devise has been incorporated into these rites of passage. (And here I'd thought Hebrew school was torture.) The idea, I guess, is that if it doesn't kill you it'll make you strong.

Two modern-day cultures can serve as examples of how puberty rites reflect the skills valued by the culture. Note that the degree of physical pain involved—whether it is a lot or a little or none at all—does not affect the validity of the rituals themselves. Puberty rites are profound, never-to-be-forgotten events.

Some Plains Indian tribes, for instance, participate in the annual sun dance, an extremely complex ceremony of several days' duration marking an adolescent's transition from childhood patterns to adult ones. Its ritual details vary from tribe to tribe, but its most dramatic element is the flesh offering. First, strong rawhide bands are attached to the top of a tall pole symbolizing the tree of life. Tribal elders incise the flesh of a young man's chest and draw a thick animal bone through the deep cut, then tie the rawhide bands to the bone. Suspended from those bands, and without using his hands, the young man must whirl around the pole until the bone tears through his flesh, disconnecting him from the pole. Only continuous pain will satisfy the ritual, until finally he breaks loose.

The breaking loose is symbolic as well as physical: he is leaving his childhood patterns behind. His fear, energy, pain, and fatigue (and, no doubt, the release of endorphins) combine to replace the old patterns with the new. Now he is a man among men, a warrior among warriors.

As Jews evolved from an aggressive, tribal culture to one of learned men, the scholar became the societal objective for Jewish men, and the pattern of the ideal male changed from warrior to intellectual: teacher, rabbi, prophet, philosopher, all of them learned men. The Jewish puberty ceremony—the Bar Mitzvah—evolved to reflect the changed pattern objective, as well. As I said earlier, the Jewish boy endures years of study in preparation for the ceremony. At thirteen, he should be able to read Hebrew and to understand the basics of the Torah and the Talmud. At thirteen, he is considered responsible for his own actions. In fact, during the ceremony the boy's father recites a blessing thanking God for relieving him of the burden of his son's sins. And at last, requirements for entry into Jewish manhood having been satisfied, there he stands: Bar Mitzvah, Son of the Commandments, ready to embrace the adult patterns of his culture.

Many cultures also celebrate corresponding puberty rites for girls in an effort to transition them into assuming the mature patterns of women. These ceremonies may resemble the Navaho girl's *kinaalda*, described earlier, or they may be something quite different. In some cultures, seclusion is an important ritual element. That is, with the onset of her first menstrual period, the girl is secluded in a small hut for two to four weeks; there, elder tribeswomen instruct her about sex, birth control, and the responsibilities of womanhood.

Some mothers greet their daughters' first menstrual period with the dictum that menstruation is a "curse," hand them some sanitary supplies, and then just leave it at that. A friend tells me that is exactly the scenario she experienced, at thirteen, with her first period. Fortunately, she already had read enough about menstruation to understand what it meant, and the day it happened, she said, she felt proud. But when she told her mother her feelings, the response was "Oh, you're just being disagreeable."

If the most important and influential person in a young girl's life tells her that menstruation is a curse, what a distressing beginning to womanhood that must be—maybe as psychologically damaging as some of the cruder female puberty rites are damaging to the body.

Menarche itself, of course, is a universal puberty event, an experience that tells a young girl definitively that she is becoming a woman. The changes involved in preparing for fertility, sexual activity, and reproduction are basic and profound. The girl's body is very quickly taking a woman's shape that, whether or not she feels ready for them, attracts boys and men. To her dismay, she may develop oily skin and acne. Her emotions are so erratic that she can hardly understand them herself. Stimulation of the ovarian and pituitary hormones is what causes all these changes that enable her to emerge from childhood's cocoon, spread her wings, and fly into maturity. The young woman's ultimate flowering, however, is accomplished with the birth of children. Motherhood is the final childhood pattern-breaker. (It is true that girls usually mature faster than boys; it seems that this maturation is nature's way of ensuring that babies will have at least one responsible parent.) Even without a community-recognized puberty ceremony, physical and psychological upheavals are sweeping the young girl toward womanhood.

If she's lucky, there will be an understanding woman to guide her on this journey.

AMERICAN ADOLESCENCE

THINGS AREN'T WHAT they used to be.

Contemporary Americans treat puberty differently than earlier generations did. We have eliminated most of the stress, fear, pain, and fatigue that once accompanied puberty ceremonies and have tried to turn our children's transition to adulthood into a walk in the park. But as any parent of teenagers knows, it isn't. Although we may have some emotional distress in preparation for a bar mitzvah or a

167

confirmation, in general our adolescents seldom face the trials of those in more tradition-bound cultures, or form the tribal bonds that mark the end of childhood. Nor, except perhaps among some Native American tribes or a few recently-immigrated groups, would we have it so.

Instead, we prolong their childhood, continuing to protect and guide and direct them through adolescence and sometimes beyond, as long as they will permit us to.

But adolescents themselves recognize a primitive and universal need to create strong emotion and high energy in order to break through into adult life. Stress is essential to generate the energies necessary to establish patterns of maturity.

Adolescents need desperately to feel that they *are* becoming adults, that they belong, that they are valued members of a group. Lacking a firm pattern of introduction to adulthood, they may join gangs ("brotherhoods" with their own rituals, rewards, and punishments); they are vulnerable to cults.

In less developed societies, adult males take responsibility for guiding their young men through rites of passage into maturity. The whole community of men are involved directly; it is their obligation to control the rituals and shape the patterns of the newly initiated. Among cultures that celebrate female rites, it is the elder women's duty to teach young girls the patterns of womanhood. In both cases, adult involvement helps ensure that the "new" men and women will continue the culture's successful patterns.

Adults in Western culture, however, seem in large part to have given up control of puberty rites for young people. Instead of taking direct part in the pattern-breaking that occurs with these rites, we expect our adolescents to stumble through, for better or worse designing their own rituals. Thus young people are left to conclude on their own what it means to be adults.

That is not necessarily a bad thing, as it may produce the kind of independent, creative people that America values. Our culture's greatest strength lies in its diversity, so that that we experience our passages alone, struggle with our demons alone, and thus arrive at a unique adulthood.

To some degree, that is.

Although most men in our society don't participate in formal puberty rituals, they apparently sense that a tribe, with its codes and secrets and shared beliefs, gives young men an identity. Therefore, men involve male adolescents in groups that act cohesively and with shared purpose—sports teams, for example, and fraternal organizations.

One result is that many men emerging in our culture have brought to adulthood a perception of the mature man as one who establishes certain kinds of independence, yet who essentially remains identified with an adolescent group—the high school football team, for instance, or the college fraternity—rather than with the culture of their fathers and grandfather, as would be true of a more evolved society.

Throughout their adult lives, these men identify closely with the mores of their youthful groups, regress rapidly to their ways of thinking and acting, and often urge their sons to join the same organizations.

In this regard, the same woman who recalled for me the day of her first menstrual period also told me another illustrative tale. For years, Elaine said, from eighth grade through college, she was deeply in love with Alan. He was her dream. At some point she and Alan parted ways and lost contact with each other, yet she continued to think very fondly of him. As the years passed, Elaine married, reared children, traveled, returned to school for a couple of graduate degrees, divorced, became a college professor. Alan remained in their home town, where he amassed a fortune in banking. Forty-five years after last seeing Alan, Elaine and he got together again.

"I was thrilled," she said. "It was only a two-day visit, and at first he was everything I'd hoped he would be—still handsome, still charming and funny. A darling, really. But before the second day was over, I could hardly wait to board the plane for home. He was such a disappointment. Charming and funny, yes, but in the same way he'd been as a teenager, except that now his wit sometimes had a mean little racist or sexist edge to it.

"He said he attends every one of our college's homecomings and class reunions. The worst thing was that he kept talking about his fraternity brothers—they haven't left the area either, and they're all still best friends. He told me they even greet each other with that dumb secret handshake frat boys learn. I guess he just hasn't grown. But good lord. The man *owns* three banks! How can he still be such a child?"

Americans' relatively undirected transition to adulthood allows flexibility, but it is also more fractal. It encourages dissident groups, a breaking away, a budding off from generation to generation and from group to group. Thus each new generation and each new group has a different understanding of the message being passed along.

Does manhood or womanhood mean this?

Or that?

Or some other thing altogether?

As modifications to the original message are themselves successively changed, social coherence breaks down to the point where it is hard to recognize the adult value patterns that originally underlay it. But as I have suggested, this may not be a bad thing. No one wants a stagnant society. On the other hand, we don't want anarchy either. A natural element of America's social patterning is the tension between our expectations for our adolescent children and their expectations for themselves. Since we have rejected the practice of forcing our children into molds, we'll have to hope that patience, understanding, and the passage of time will resolve it.

Or maybe not. It seems to me that what is starting to evolve now is the possibility of universal rites, i.e., developing entirely new beliefs and practices as information travels instantly and continuously to every corner of the earth. It is interesting to imagine what puberty rituals, adult societal beliefs, and global attitudes may emerge in this era.

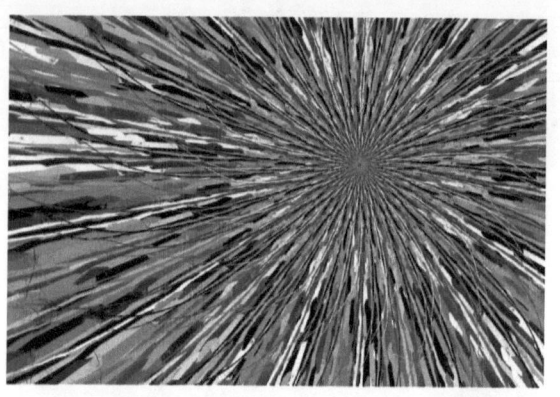

EMOTIONAL ENERGY
AND PATTERN CHANGES

UNTIL DISNEY GOT hold of them, the stories we now call fairy tales were never intended specifically for children. Most of these ancient folktales originated ages ago in the dark, forbidding forests of northern and central Europe. Replete with violence, death, magic, good and evil, crime and punishment, they are morality tales, cautionary tales, and just plain horror stories.

174

Today you seldom find them on the children's shelves at bookstores, as they are considered too unsettling; most children are familiar with a few of them only through the bland Disney versions.

Without strong models to guide them, adolescents find their own ways to generate the emotional energy needed to make the pattern changes that allow them to grow up. They seek thrills.

Perhaps you have wondered why so many teenagers enjoy slasher movies—the bloodier the better. These films typically involve a psychopath stalking and killing a sequence of victims in luridly graphic ways; often the victims are adolescents, and usually the audience—mostly teenagers who have outgrown Disney—follows the action, including the vicious cutting and slashing, through the killer's eyes. Usually, good triumphs over evil; sometimes, however, like Freddy Krueger of the *Nightmare on Elm Street* series, the villain re-emerges in multiple sequels—a chilling reminder to the young audience, perhaps, that evil always lurks.

Slasher movies offer a safe thrill. Other thrill-seeking activities are much less safe—alcohol and drugs, for instance. Young people, vulnerable to peer pressure and thinking they

175

are immortal, will try anything once—drugs, alcohol, any substance that promises nirvana—never for a moment believing the disastrous results that happen to other people can happen to them. Too often, moreover, that first "once" turns into habitual behavior, which is physically, psychologically, and financially costly.

Teenagers probably consider sex the ultimate adult pastime.

Why wouldn't they?

Books and movies and TV have been telling them so ever since they were born. Yet early sexual experimentation can have dire consequences. Probably those scary news reports about sex among preteens inflate the numbers somewhat, but there is no denying that it happens. And sexual relationships among older adolescents seem to have become the order of the day.

Many—most?—parents find this unacceptable, yet they have no realistic choice but to accept it; short of chaining them to the wall, there is a point beyond which you can't control your children's behavior. Preaching abstinence to hormone-driven adolescents is a waste of breath. No more than twenty or so years back, we were startled to hear about mothers who put their daughters on the pill as soon as they began menstruating. HPV vaccination is recommended by the Centers for Disease Control and Prevention for preteen girls and boys at age 11 or 12 years. Some doctors recommend it for girls as young as nine. Why so young? Because the highly prevalent human papilloma virus responsible for genital warts and cervical cancer is sexually transmitted, girls need to have the vaccination before they become sexually active.

Patterns of sexual behavior among adolescents have changed markedly in the past century, at least in the sense that there is a whole lot more of it, and societal attitudes regarding it have changed too, albeit more slowly. It is important is to equip our children with real, useful knowledge.

In addition to learning the mechanics of sex and contraception, they need to understand how young women's sexuality differs from young men's; this may help them avoid that series of misinterpretations that can culminate in what is known as date rape, or, at best, in bruised feelings. We can't, certainly, prepare our children to deal successfully with all the hazards inherent in human sexuality, but it is useful to remember that if they have been reared with love and respect, they are most likely to treat others the same way.

Gang membership offers another way to stir up the energy needed to break early patterns. Creating a gang, a substitute family, is a common phenomenon. Like a team, a gang fosters group identification and spirit, but the identification is more impassioned and gang activities are more dangerous. Gangs are rare in cultures with pattern-changing puberty rites, although they frequently emerge as these cultures deteriorate. But when young males in our society form gangs, they are saying, "If none of you adults will help bring us into maturity, then we'll do it ourselves." Thus the gang replaces the adult male tribe.

Teen gangs adopt branding that symbolizes membership: distinctive clothing, colors, hairstyles, tattoos, and so on. The organization is tight—constricting, really, because for these young people the world has grown small— and identification with the gang becomes increasingly intense, so that the greatest sin is defying the group. Usually these young men lack strong family structure; the gang *is* their family: their mother, father, community, and country. In large cities such as Los Angeles and New York, sometimes

178

law enforcement finds itself negotiating with powerful gangs as they do with other factions in our culture.

For all its negative characteristics, gang behavior tells us that young men need creative ways to escape the constraints of childhood into the freedoms of maturity. To address these issues of adolescent rebellion and to steer our young people away from danger, we must provide them with rites of passage—experiences that test their mettle. They should expect some wrenching emotion, some suffering, and some failure in order, finally, to be accepted, and to know themselves, as men.

THE PERILS
OF OVERPROTECTION

PAUL EHRLICH DEVELOPED the world's first chemotherapeutic agent for successful systemic treatment of a micro-organic infection, namely syphilis. Ehrlich called the compound Salvarsan 606. Before Salvarsan 606, he had developed and tested 605 compounds that failed to cure the disease. Later, he produced an improved version: Salvarsan 914. Not only was Ehrlich unintimidated by all those failures, he didn't see them as a reason to stop trying; moreover, he learned something from each of them that spurred him to press on.

Overprotection is oppressive. Children who are overprotected become afraid to fail and will do anything to save themselves from failing. When a child's life is too easy and he doesn't have to struggle, it's hard for him to take risks. He hangs back fearfully while other children forge ahead of him, happily learning to ride a bike, swim, do long division and calculus, write books and maybe find a cure for cancer. Instead of teaching children that success is the norm, we should teach them to fail at their highest level, assess the situation, and try again.

In the earlier days of this country, children had valuable work to do. On farms, they were expected to attend to their chores, go to school, learn their lessons, and if there was time, play. They worked the fields beside their parents. Like the adults, they fought flood and drought. They suffered through epidemics, often helping to bury their sisters and brothers, and sometimes their mother and father.

Many grew up fast, and because they had learned competence, they grew up self-confident.

So too did the children of poor immigrants learn to take care of themselves, work, contribute to the family purse, and shoulder serious responsibilities. The conditions they faced made these children strong, tough, and fearless. My father was a child of those times. The son of nearly penniless immigrants, as a boy he sold used shoes on the streets of Columbus, Ohio as a means to help keep food on the family table.

Dad graduated from hawking old shoes to running a cobbler's shop, from there to opening a shoe store, and after that the sky was the limit: he established one of the largest, most successful shoe businesses in the United States and took his company all the way to membership in the New York Stock Exchange. Dad was smart, competent, game to take risks.

And nobody told him not to.

Once a young person is trapped in a pattern of values that says he should never fail, he begins to truncate his goals, limiting himself to what is safe and certain and thereby undermining his ability to realize his full potential. If, however, he is allowed to struggle and strive, extending his energies to the point of failure, he then can set his goals well beyond his current—and temporary—ability to achieve. This much healthier, more creative mindset encourages continued effort and continued achievement. After all, as Browning tells us, a man's reach must exceed his grasp, or what's a heaven for?

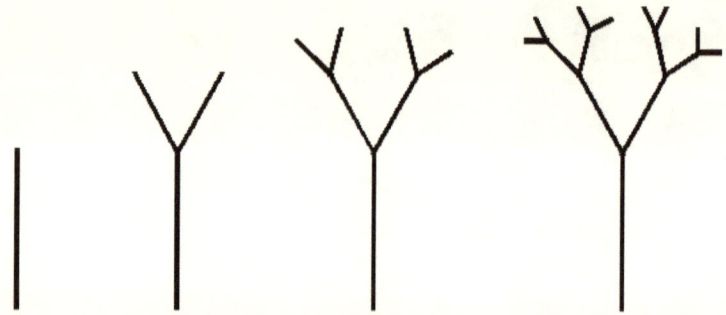

ADOLESCENT PATTERNING
AND COMMUNICATION
TECHNOLOGY

ALMOST FROM BIRTH, today's adolescents have been children of technology. Even if their parents managed to limit their hours of TV watching and computer use during the

184

first few years, by the time they are teenagers it is nearly impossible to impose such restrictions.

For one thing, there is immense peer pressure: "But Mom, *everybody* has a cell phone." Or a Twitter account, or a Facebook membership, or an iPhone or iPad or whatever is the latest product of communication genius. And it's true; almost everybody—including Mom herself—does have one or several of these items.

Furthermore, nearly all of us have come to appreciate their usefulness and convenience. Say, for instance, you bought a cheap cell phone a few years ago "for emergencies"—maybe a breakdown on the highway. Since then, however, your phone (that is, your newer, finer phone with its infinity of apps) has become your constant companion. At this point in our developing technology, it just seems wrong-headed, not to say awfully stodgy, to deny these marvels to our teenagers. And what a welcome surprise it has been to learn that the devices can actually help us stay in contact with them.

Yet there are pitfalls.

We are all aware of the very real dangers to our children of internet predators. We know about "sexting." We have heard the news stories of young people's lives shattered by the concerted efforts of ill-wishers to destroy them by circulating scabrous comment through Facebook. Such activities are deplorable, of course, and very difficult to circumvent; perhaps the best we can do to protect our children is to be sure they are well informed about the perils of electronic communication, and then hope that we have endowed them with sufficiently strong values to keep them out of trouble. (It doesn't make our task any easier that even some members of congress have engaged in the very practices we warn our children to avoid.)

In halcyon days of yore, there weren't any teenagers. Oh, there were plenty of adolescents, but the word "teenager" didn't come into common usage until late in the second half of the twentieth century. Before then, adolescent children ("youths") were expected to obey their parents' rules, respect their elders, and comport themselves like grown-ups. Pretty much, they did—with allowances made for their immaturity. Gradually, however, enough information from child development experts reached the general public to cause parents to start considering their adolescent children as a separate species. Today we even acknowledge adolescent sub-groups. There are (depending on whom you ask) preteens (eight to ten), tweeners (ten to twelve), young teens (thirteen to fifteen), and young adults (sixteen to eighteen and older). American businesses, such as toymakers, electronics companies, clothing manufacturers, and publishing houses, have adopted the classifications enthusiastically, tailoring and advertising their products to appeal to each age group.

One result is that the children themselves accept the various labels and begin to self-identify according this new taxonomy. Abetted by the social networking culture, they distance themselves farther and farther from the world of their parents and into a world of their own.

Using technology to communicate throughout the planet, young people are sending and receiving ideas and values vastly different from those of their family and community and country.

This international impact can have positive results. For example, led by young people with cell phones, a new era has begun in the Middle East, with dictatorships toppling one after another. It was called the Arab Spring.

What an apt term. Youth is the springtime of life, when anything seems possible, including breaking away from oppressive rule to seek self-determination. You may recall that during the demonstrations that resulted in Hosni Mubarak's removal from power in Egypt, Mubarak addressed the revolutionaries as "my children." As he learned to his chagrin, it was exactly the wrong thing to say, at exactly the wrong moment.

Social networking may promise a more peaceful world now that children can learn first-hand to appreciate other cultures; perhaps, newly empowered by technology, eighteen- and nineteen-year-olds will resist going to war against people they have "friended." On the other hand, though, chaos is a possibility, at least in the near term. Social networking among the young can undo patterns that have been in place for centuries, and young people, as members of a world culture, will have to supplant them with their own.

But think back.

Technology always has been a change agent for social patterns. Five-hundred years ago the invention of movable type made books—and all the new ideas they contained—readily available to everyone who wanted to read them. The industrial revolution propelled huge migrations of people from farms to cities, where they learned new patterns of thinking and living. The mass production of automobiles set young Americans on the move; they left the family farm or the small town where they grew up, and like Huckleberry Finn, "lit out for the Territory"—new places where they might be free to establish new patterns. It is only to be expected that today's extraordinary technology will produce major pattern change. The difference is that it is happening so fast. Change alarms people, and rapid change terrifies them.

Depending on how we approach it, what is happening in this swiftly evolving universe of instant communication and social networking can redound to our benefit, or further splinter our culture. First, it is well to remember that among human beings there is a genetic constant that demands patterns of love and caring, so that when the technology revolution has stabilized, those essential patterns will still be in place. Though the forms of family and society may be different, their substance, of human necessity, will remain.

But you can't stop the revolution. You can choose merely to stay calm in the face of it. You can dismiss it, become a member of the Society of Obstinate Coots—and thereby widen the gulf between yourself and your children.

189

Or you can join it. Bringing to bear the experience and insight of maturity, perhaps, like Ben Franklin during America's revolutionary years, you'll be able to temper the excesses of youth with the wisdom of compromise. Yet even if you don't aspire to make history yourself, armed with minimal technical expertise, you will be part of a movement that, by sowing information throughout our world, in time may empower nearly seven billion people to enrich the quality of their lives.

PATTERNS IN ADULTHOOD

FIVE-THIRTY, FRIDAY AFTERNOON. Man walks into favorite bar, sits on customary bar stool.

Bartender asks, "What can I get you?"

"The usual. Hey, where's Joe?"

"He quit yesterday. What can I get you?"

"Joe quit? Why? Where'd he go?"

"No idea. Sir, what can I get you?"

"Uh, yeah, Scotch. On the rocks."

"Any preference?"

"Walker, Black."

Man's nonplussed. *What the hell happened to Joe?* Man's been coming to this bar practically every Friday afternoon for six years. Joe's been a fixture. They'd trade a few jokes, complain about their wives, rag on their dumb-ass bosses—talk about damn anything at all.

Man'd have a couple of drinks, no more, no less—just a way to say so long work, hello weekend.

Now what?

This kid behind the bar. Looks about fourteen. Hair sticking up in orange spikes. Gold rings in both ears. Gay? Who knows? Man used to be able to tell. Can't anymore. All these kids look alike. Man knows one thing though: no more nice, easy conversations about wives and bosses, not with this freaky kid.

Where the hell did Joe *go*? *Why?*

Man has weird sensation life has just lurched sideways.

From the bath, story, and bed routine that sends your toddler securely into sleep, to the rituals of extreme unction that ease a terminally sick man's passage into death, people need patterns. Imposed by others or generated from within, patterns help us understand and control our world—or, sometimes, hinder our understanding and control. In adulthood, we see the fullest expression of many patterns that originated in childhood and adolescence. Genetic and perceptual patterns, honed by years of repetition, have become entrenched.

By the time people reach adulthood, they have learned and stored huge numbers of patterns that have become increasingly efficient–so efficient, in fact, that a person can go through a whole day exercising virtually no independent thought. "Get up, get out of bed, drag a comb across your head," as the Beatles put it, and as the hours pass we unconsciously summon up a host of other patterns, one after another, with which to respond to whatever the day has in store for us. If our patterned responses are healthy, we will use our energy for useful purposes. If they are not, there will be trouble.

Take, for example, that fellow in the bar. His pal the bartender has disappeared, probably for good. His Friday-afternoon pattern has been breached and he's feeling betrayed and very glum. Many choices now confront him. He can, for instance, find another bar, with maybe an equally convivial bartender. He can stop after one drink, go home, complain to his wife about his boss, and take her out to dinner and a movie. He can choose to really tie one on, try to drive home, and end up calling his wife to bail him out of the drunk tank.

How he decides to handle the unexpected problem of his vanished bartender depends in large part on how he has been patterned–by genetic, social, and self-imposed forces.

Maturity in humans is admittedly an ambiguous quantity. Who among us, having smugly assumed he has achieved maturity of thought and action, has not done childish and ill-considered things that undermine his sanguine self-assessment? I certainly have, and some of those things are still too embarrassing to dwell on. But for the purposes of this book, I wish to define maturity as that point in people's lives when they have developed their own patterns of belief–patterns that correspond to what they, personally, know to be true, rather than to what some anonymous "they" believe.

"*What will they think?*" is an override that curtails the ability to think for yourself and stand behind what you believe. This barrier to independent thought arises naturally because in childhood and adolescence what others think is crucial to development and acceptance. First, the family and the community program the child, helping him shape his behavior until it meets societal expectations. In adolescence, a different tribe–the young person's peers–becomes the formative matrix, and what is acceptable to the peer group is likely to diverge wildly from what the family expects. Not that the family and community's influence has waned: its continuing force is made obvious by the adolescent's mighty efforts to break away from it. Your honor-student son, for example, is caught smoking in the high school parking lot and is suspended from school for two weeks. Your beautiful sixteen-year-old daughter comes home from a trip to the mall sporting a nose ring and a tattoo. It is hard for most parents to consider such maddening and ludicrous behaviors as signs of developing maturity, but that indeed is what they are.

Throughout adulthood, of course, we go on joining other peer groups–social groups, business groups, religious groups, and so on. The trick, however, is to avoid letting group-think supplant independent judgment, and sometimes that can be as difficult for an adult as it is for a teenager. My own political evolution is a case in point. For many years I was a pretty staunch Republican, as were most of my friends and associates. I believed that if a party based its political philosophy on fiscal responsibility and social progress, then it was the party for me, and that what was good for Ivan Gilbert, prosperous businessman, was good for the country.

But some years back I began to see what I believed to be a radical change in the Republican party, that it had become the party of fiscal *ir*responsibility and that it not only opposed social progress but was fearful of it. So for me to continue my Republican allegiance now created a very uncomfortable cognitive dissonance, which I resolved by switching parties. The surprise was this: my friends stayed with the Republicans. Why? They're smart people, and what was obvious to me must have been equally obvious to them, so why, instead of acknowledging the logic of what I had done and following suit, did they remain obdurately Republican and see me as a rather muddle-brained defector? I can suppose only that escaping their long-held political pattern was too challenging, perhaps too frightening, for them.

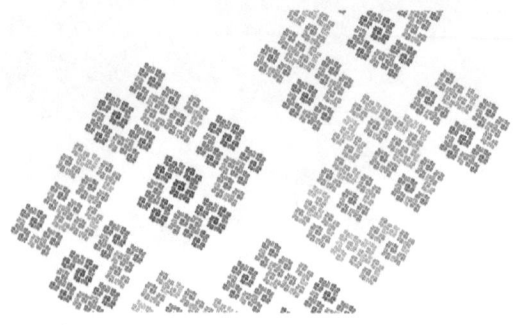

INDIVIDUAL RIGHTS
AND
SOCIETY'S RIGHTS

TO BE ADULT is to be willing to defy the group and establish independent beliefs. But while jettisoning stale and counterproductive patterns and creating a strong personal identity, it is important to remember that we are also members of a wider society, and that we have responsibilities to various groups: our families, our institutions, our nation, and even to the world itself.

In groups with fewer resources, survival requires that people subjugate their self-interest to the needs of the tribe. But increasing affluence creates the luxury of individualism and a correspondingly increasing tension between self-interest and the common interest. Our very wealthy culture, for example, is proud of its individualism; to some, the successful "self-made man" represents the finest harvest of the American grain, the acme of individualist thought and achievement. The individualist worldview permeates our history, our literature, our science—indeed, it informs the whole of America's five-hundred-year-old mythology. Don't tread on me. Live free or die. I know my rights.

But much of this, as the preacher saith, is vanity. For the thing that so many adherents of this worldview fail to appreciate is that it is just fine as long as it doesn't disturb the group's equilibrium. Realistically, the rights of the individual are like a little boat floating on calm waters. It will remain afloat as long as the group that buoys it remains calm. But when a hurricane hits, the boat may be smashed to pieces. Individual rights exist only when the group chooses to allow them to exist.

For example, the mid-twentieth century saw individual rights stripped away when the American body politic became infected with a fear of Communism. The

House Un-American Activities Committee and Senator Joe McCarthy, with help from the FBI, continually and with impunity violated the rights of individual citizens. Today we look back at this period in our history and shake our heads in dismay at the thought of so many people's lives having been destroyed by demagoguery. Yet have we really learned anything?

After the September 11, 2001 terrorist attacks, despite the objections of civil libertarians, Congress enacted the fear-driven Patriot Act, and the Bush administration frightened America into war on Iraq, a country that had absolutely no part in the 9/11 attacks, and impugned as "unpatriotic" the motives of those independent thinkers who were against the war.

The fragility of individualism becomes even more clear on the quotidian level. Suppose it's winter and there is a massive failure of the electric, gas, and water systems. Ice has downed telephone lines. How long could people survive in such a predicament?

Not very long.

They would realize quite quickly how dependent they are on the efforts of myriad other people to repair and maintain those systems. Recently, a heavy snowfall hit New York City and within a couple of days, because city workers had failed to clear all the streets, New Yorkers were calling for the mayor's head.

Once, Washington, D.C., our center of government and a city that cowers under a half-inch of snow, experienced a storm that left three feet of snow in its wake. The bureaucracy shut down and went home, and so did the Congress and the Supreme Court. (The President was already at home in the White House, so he had no excuse to flee.) Nobody came back to work for a week. For an entire week, then, while an army of workers struggled to clear the roads and bridges into Washington, American government almost ceased altogether.

So much for rugged American individualism.

I don't mean to suggest that the spirit of independence should not be fundamental to our culture. It is worrisome, though, that in our growing focus on individual entitlement we may have lost sight of family and community entitlement. We need balance.

Most men and women today accept that they are fully equal, independently functioning human beings. And that is a good thing, unless it is taken too far and people choose to appease their own ambitions and needs rather than those of, say, the family.

It is inarguable that we are seeing a far too casual breakup of families (half of first marriages in America end in divorce), in effect the interchanging of parts, so that children are apt to have four or six or even more parents during their lifetime. Such children will likely have great difficulty establishing rewarding relationships; after all, if mommies and daddies are replaceable, so are friends and husbands and wives.

They grow up replicating their parents, insisting on their individual entitlements instead of trying to build relationships. And further to exacerbate the disintegration of family structures, many of these children will use the services of on-line dating sites (at least two of which now are matching people seeking partners for extramarital affairs).

Some sociologists warn that internet dating services "are gradually changing society's conception of relationships and marriage by encouraging people to view partners as commodities that can simply be traded in for better versions at the click of a computer mouse."

Moreover, this self-serving pattern has extended through the whole of our society. Too often we see ourselves as entitled to much more than our share of the world's resources, so we gobble them up and spew out their residue, giving insufficient attention to the needs of others and to the planet itself. If strong human relationships are the core of a strong society, then it must be true that without them, as Yeats tells us, things fall apart; the center cannot hold.

This pattern we have established, in which personal independence is key, is a virus transmitted through generations, and perhaps only a cataclysmic event will demand that we seek a cure. I recall the World War II years clearly. For instance, there was a time when Americans willingly sacrificed for the good of other peoples, for the common cause.

And we all remember how in the days following 9/11 we came together as Americans first, and black or white, Jew or Gentile, Democrat or Republican. Sooner or later some other earthshaking disaster will occur. The possibility of nuclear war, for example, is always at the edge of our consciousness. At such a time we will return–because chaos is the only other choice–to the old belief that the group must survive before the individualist can claim any rights at all.

THE PATTERN OF
SUCCESSFUL ADULTS

LIKE OTHER ORGANISMS, human beings are subjects to Darwinian winnowing. Those adults whose patterns of perception and action enable them to deal effectively with their environment are the survivors.

204

Once an organism begins to succeed, it will do so repetitively, and unless the organism or its environment changes, it will continue to be successful most of the time. By the same token, organisms that do not manage stimuli as well, but still survive, will continuously fail to manage well, and will continuously be less successful.

Their patterns, while not entirely destructive, are less adaptive to their environment. These observations are not set in stone, of course, as there are many levels and definitions of success. But so much for organisms. Let's get down to cases.

First the bad news. Not everybody succeeds. Unforeseen events, ill-timed decisions, bad judgment, personal weaknesses—a world of trouble awaits the would-be success. A person can even become prey to his own success. Say, for example, that you're a successful restaurateur whose life's dream has been to operate a fine seafood restaurant in New Orleans. It will be a costly venture, but you have been prudent, and you have amassed enough capital to do it. You buy the building; the location is perfect. You engage a peerless chef. Your restaurant—call it *Le Bon Poisson*—is scheduled to open tomorrow night. Today, however, a huge oil rig explodes in the Gulf. Millions of gallons of oil gush from the well, suffocating marine life, blighting the beaches, and destroying your dream.

Or perhaps you invested in Lehman Brothers the week before it collapsed. Or turned your life savings over to Bernie Madoff because everybody else was doing it, and gosh, the guy seemed like such a mensch. Or gambled away your child's college fund. Or you fell victim to the Peter Principle, a theory proposed in 1969 by Dr. Laurence J. Peter. His theory holds that employees within a hierarchical organization will rise to their highest level of competence and then be promoted to and remain at a level at which they are incompetent.

Consider, for instance, Ms. Willison, a skilled, compassionate third-grade teacher. She doesn't merely educate her pupils; she teaches them to pursue the joy of learning. They love her, and so do their parents. Her faculty colleagues, moreover, consult her for guidance and leadership. In addition, she is admired for her meticulous records-keeping.

Ms. Willison, therefore, is the obvious choice for elementary-school principal, a job in which she is so successful that within a couple of years she becomes head of a middle school. She finds more challenges in this position, but after a somewhat rocky period of adjustment she rises to them. Then a high-school principal position opens. Ms. Willison is promoted. A host of new problems now confront her: ragingly hormonal adolescents, a fractious faculty, irate parents, the impossible demands of No Child Left Behind.

She feels besieged; her experience hasn't prepared her for all this; her skills and personality are a poor fit for this position; she simply isn't up to the job. But there, without prospect of further advancement, she will remain. Ms. Willison has risen to her level of incompetence.

And yet, despite the many barriers to success that face us, there are people whose every venture seems to turn to gold. We marvel at their success, at their soaring trajectory. Luck may play a part in their success, but usually it is only a minor role, for such people are those whose patterns are so well adapted to their environment that only a catastrophe, such as that oil rig explosion in the Gulf of Mexico, would cause failure of a new project.

In every aspect of a society we see the early rise of the successful person, who then continues to repeat his successes throughout his life. And by the same token, people who are less well adapted, but still survive, will continue to be less adaptive and will continue to be less successful.

Certainly I don't mean to suggest that success is a matter of arbitrary definition, of either achievement or failure, as success is a nuanced entity. A person may be quite successful in one environment because his patterned responses are well adapted to it, but be less successful in another environment because the very patterns that worked so well in the first place are maladaptive in the second. The unfortunate Ms. Willison found that out.

As another example, probably all of us can point to extremely successful professionals who are incompetent parents or spouses; the marketplace-patterned skills that serve them so well in business may be dysfunctional in family relationships. And the child psychologist who can't manage her own children, the *haut cuisinier* whose personal diet preferences run to Big Macs and Pepsi, the brilliant satirist who can't take a joke—such figures are the stuff of cliché.

It is important to know that patterns of success and patterns of failure have relatively little to do with that ingredient we call intelligence. Many highly intelligent people fail; others who appear less intelligent can be highly successful. The latter often operate on an emotional, intuitive basis; they seem able to read their environments and adapt quickly, and this ability probably has little to do with intellect *per se*.

Such people's patterns may seem aberrant. They'll do things that look unconventional, but that is because they have sensed a need for change, repatterned themselves, and adapted more quickly than those who stick with more conventional patterns. Professional boxing is illustrative of what I mean. To prevail, the boxer must not only bring all his training, conditioning, skills, and strength to the match, but he must also read his opponent and adjust in split seconds to what he reads so that he knows whether to float like a butterfly or sting like a bee. And speaking of Muhammad Ali, once, at a pre-fight press conference, a reporter brought up the matter of Ali's IQ, which had been measured at 85—well below average. Memorably and without missing a beat, the boxer replied, "I never said I was the smartest. I said I was the greatest."

Some areas of successful functioning, such as notable skill in the arts, finance, and politics, are obvious to everyone. Slip-ups and failures among the celebrated are equally obvious. And the observer's perspective, too, is based on patterned perceptions and subjective beliefs and feelings. The highest-profile job in our time, for example, is the American Presidency.

Less subjective, though, is the case of the born leader: someone who arrives on this planet blessed. His or her effectiveness can be modified or misdirected by environment, of course, yet the basic ability to lead is found in the genes. Men and women bearing this genetic gift are perceived more easily as leaders, and they are seen as sexually attractive. Underlying courtship and romance is an instinctual drive toward survival of the species; at an unconscious level all people want to mate with a successful person so that the superior genetic material will contribute to superior progeny.

People identify attractiveness and sexual appeal based on their society's patterned values. Different societies define winners in different ways, but a woman's choice of a strong, masculine man and a man's choice of a strong, feminine woman are much more than culturally derived patterns; they are a matter of survival. Success breeds success—literally.

Henry VIII and his daughter Elizabeth I come to mind. Henry was attractive, educated, accomplished, charismatic—and according to reports a very sexy fellow. In order to divorce Catherine, his first wife, who had failed to provide a male heir, Henry broke with the Papacy and named himself Supreme Head of the Church of England. The majority of the English populace accepted this bold move; with its impetus, Catholic England joined the Protestant Reformation.

Henry married his second wife, Anne Boleyn, with whom he was besotted and who he hoped would provide him a son. It did not happen, the marriage was unhappy, and by order of the King, Anne was beheaded, leaving Henry to pursue his quest for a son with several more wives. Anne did, however, leave behind a little daughter: Elizabeth.

That bare-bones account of Henry VIII's life is familiar to all of us. What is less familiar is the daunting tangle of personal and political relationships, alliances, and intrigues through which the King must make his way in order to rule successfully. Only a born leader could have done so, and Henry was that leader. By the time of his death, circumstances had combined to make Elizabeth heir to his throne.

Henry had always believed that no woman could manage a kingdom, but his daughter proved him mistaken. As attractive, accomplished, and charismatic as her father, and forced to navigate an even more turbulent sea of relationships and intrigues, Queen Elizabeth I guided her country through its brilliant High Renaissance, a period during which England achieved supremacy in military affairs, politics, exploration, science, medicine, literature and the arts.

Despite being courted by princes and wooed by many less illustrious suitors, Elizabeth never married— possibly because her father's treatment of her mother had left her with a dim view of marriage, but more likely because she, like Henry, was above all things a canny politician. She must have sensed that sharing her personal or political power with another crowned head might dilute her strength of leadership. She chose not to marry, but to lead.

Royal birth may have given both Henry and Elizabeth an advantage from the start, but without immense strength of character, will, and vision, they both would have joined the long line of English monarchs who, in most minds, have faded indistinguishably into history.

It is also true that people from much humbler beginnings can achieve spectacular success. Richard Branson, seriously dyslexic, was an academic failure, bouncing from school to school until he finally managed to graduate. Yet beginning with his publication of a small school magazine, Branson's leadership skill and business acumen led him to establish Virgin Records, which developed into the multifaceted, highly successful Virgin Group. His best known venture is Virgin Galactic, a commercial space travel company. "My interest in life," Branson says, "comes from setting myself huge, unbelievable challenges and then trying to rise above them."

And the late Steve Jobs, founder of Apple Computer, didn't bother to finish high school. A visionary with extraordinary drive, Jobs was able to see far beyond anything further schooling might offer him; his work has revolutionized communications technology, and the end of the revolution is by no means in sight. In no small part, Jobs's genius is responsible for the still-evolving changes in how people view themselves and their countries in relation to the rest of the world. How did all this come about?

Jobs liked to quote hockey great Wayne Gretzky: "I skate to where the puck is going to be, not where it has been."

WOMEN

QUESTION: HOW MANY men does it take to engender a hundred children? Answer: One, if he keeps fairly busy. Question: How many women does it take to bear, nurture, and rear a hundred children? Answer: I'm not sure, but it's a whole lot more than one.

Throughout our species' several million years of genetic development and survival of the fittest, the groups that prosper have been those with strong, nurture-patterned women who are able to transmit love and caring to children, and who have been supported by strong men

214

who, individually and collectively, are patterned to protect and provide for the women and children. Although admittedly there are exceptions to these long-established patterns of living, almost always it is mothers who nurture the next generation, and the next and the next; if in our haste to avoid sexism and ensure that every human pursuit is equally available to every person, we ignore that fact so basic to our common humanity and we put our posterity at risk.

Let's look at a case in point.

The infamous Siege of Leningrad, which took place between 1941 and 1944, destroyed nearly the entire adult male population of the city. Only ten percent of the men survived. Yet within twenty-five years after the siege was lifted, the population was nearly identical to what it had been before the war. If the situation had been reversed, and ninety percent of the women had been wiped out, Leningrad probably would have perished. There were multiple reasons for the population's resurgence, but two in particular stand out.

For one thing, survivors of a cataclysmic event wish to celebrate, to share their joy in having stayed alive, to pick up the pieces and go on. For a while at least, barriers of social class and position fall, and a happily promiscuous mingling takes their place. The surviving women of

Leningrad were encouraged to mingle quite a lot; they did, and quite a lot of babies resulted. But another, more crucial cause underlay Leningrad's renascence: the millions-of-years-old pattern, ingrained, unconsidered in the heat of a sexual moment, that dictates survival of the species; that is, for a society to function and develop, its women must bear and nurture children.

Women are key to any society. Something that all women know but that, historically, men have been reluctant to acknowledge is this: societies really don't require many men to keep themselves going, but without women they can't survive. It follows that the protection of women is central to the continuation of a culture. Moreover, women deserve not only protection, but genuine, sincere appreciation for their enormous contribution. I believe that our society, as well as many others, has failed women in that regard. Women know that, too, and they resent it.

Roses on Mother's Day are nice, but as any mother will tell you, *every* day is Mother's Day.

Yet, in their understandable enthusiasm to achieve equal rights, many women are overlooking their pivotal generational role. In recent years, for example, American women have been authorized to serve in combat areas during times of war. As of today, women are still barred from U.S. combat groups, but in modern warfare any location in a battle area is subject to attack. In Afghanistan, the " battlefield" is everywhere. Considerable pressure has been applied to persuade our military leaders to bring about the reality of women joining combat—much of the pressure having come from women themselves. It is true that combat experience increases a soldier's chances of promotion; it also increases his chances of death.

If a woman chooses a military career, she expects it to include promotion, so that it seems logical that she demand the right to combat experience, which includes, of course, the right to be killed. But a more pressing logic warns us that placing women in mortal peril endangers the very society the military is trying to defend. It's a conundrum. Should the United States ever determine to reinstate the draft, it will likely include young women as well as young men, and given the ways of modern warfare and the existence of modern America's many equal rights statutes, the effect on our population could be devastating.

In other settings, too, women are moving away from their role as nurturers. Particularly since the last quarter of the 20th century, a cultural sense has developed that tells women that remaining at home to care for a family is a waste of their talents. In the fifties, when large numbers of young women began enrolling in colleges, the popular canard among men was that "the girls" were going to school only to earn their MRS degree. Not very funny, but in fact during those early years most female college graduates did marry, have babies, and stay home to raise them. It was the cultural expectation.

In the abstract, it is perfectly possible that stay-at-home mothers have the most creative lives of all. From raw material, her children, the homemaker molds mature, loving, competent, confident human beings who can replace their preceding generation and maintain the quality of their society. The day-to-day reality of child-rearing, however, may present a different picture. Imagine a young mother in, say, 1958. Her college education has equipped her with skills and ambition.

It has failed, though, to teach her how to use those skills, realize her ambition, and nurture a family at the same time. She loves her children very much and is diligent in their care. But her days are filled with endless household tasks, children's squabbles to mediate, runny noses that must be mopped, spilled milk or juice or Cheerios to clean up, baths to be given, breakfasts, lunches, dinners to prepare again and again and again.

In the past decade or so, there have been signs of what may be reason for hope. For one thing, some working mothers appear to be rethinking their priorities. Exhausted, unhappy with daycare, missing out on so much of their children's development, they are putting their careers on hold and returning home. They have determined that they will live with less, but live better. It is a move that will work, of course, only if there is a father working to support the family. More and more these days, many young men and women seem to regard marriage as a rather quaint relic.

In our cyber-connected world, more and more people are working from home; they and the businesses that employ them are finding the arrangement mutually beneficial. No daycare fees, no transportation and parking costs, and a life-enhancing sense of autonomy for the at-home worker.

Back at the office, fewer employees, fewer arguments among employees that the manager must referee, and—who knows?—maybe smaller and more cost-effective office space. It may sound utopian, but with cooperation and common sense it could come to pass. After all, for decades we have heard promises that the computer will create for us more productive, more convenient, easier and happier lives. It's high time it did. And if that ever happens, the office itself—that mysterious, faraway place "where Mommy and Daddy work"—may become almost a thing of the past. Interestingly, its replacement will be a very old societal pattern: the parents working at home and the children growing up there, observing and learning their parents' trade, understanding at first hand the value of work, and maturing into responsible adults.

The mother's love and discipline of her children are crucial to prevent the injection of a harmful pattern virus: one generation of inadequately mothered children can cause a society's downfall. Such children become adults who are insecure, immature, selfish, and unable to delay gratification. As they reproduce, they pass the pattern virus on to their children, and the patterns that are thus transmitted portend weakening and possible collapse of a society. A country whose citizens are fighting constantly with each other rather than building their nation together is ripe for destruction.

We see that now in the United States as our politics become deeply polarized. We have splintered into racial groups, hyphenated ethnic groups, socioeconomic groups, sexual-preference groups, rights-to-this-or-that groups, all of which view the world through the narrow lens of their particular interest. If seeking the common good interferes with one's personal sphere of concern or pleasure, it is not a priority for most Americans. During the past several years, much of the action—or inaction—of our Congress, whose membership for better or worse does represent its American constituency, is a result of this factionalism.

The family is humanity's central social unit, and throughout our history it has fallen to the mother to hold this unit together—a demanding responsibility under the best of circumstances, its demands exacerbated when the mother works away from home. With today's radical shifts in our ways of living and the pernicious pattern virus that is undermining our sense of the common good, factionalism begins at home. If the center cannot hold, then, as Yeats wrote, "mere anarchy is loosed upon the world."

MEN

IN PART BECAUSE of effective birth control methods that prevent unwanted pregnancy, thereby freeing women to choose options other than motherhood, and in part because of evolving social change, today we see young women delaying marriage and children, sometimes until very late into their childbearing years. At the same time, this new reality has affected young men.

A man's traditional role has been to marry and accept responsibility for the sustenance and protection of his family. Today, though, many young men, having been patterned for individualism, are not prepared or willing to assume this responsibility. In fact, if indeed they ever get around to marrying, they expect a wife not only to mother the children but also to carry part of the load that previously was the father's role; they won't settle for less. As we have seen, this arrangement can be exceedingly difficult for a woman, and women are right to be leery of taking it on.

And then there is the question of ever marrying at all. "*He can't commit!*" cries the exasperated woman who, ready to move from courtship to marriage, discovers that the man of her dreams is reluctant to move with her. "He's selfish. He's immature. And worst of all, he never really loved me."

Some men can commit, however, and those who stay in relationships with women during the child-rearing period develop a strong sense of "ownership"—not in the sense of having a slave, but of having a wife who is fully devoted to him and to whom he is fully devoted. Males of many species have a need for this kind of "possession" and an instinctual willingness to fight and even die to protect it.

Some women mistake this natural impulse for control or jealousy. Yet the self-confident man is neither jealous nor controlling; he simply expects an exclusive marital relationship.

It is a genetic pattern of behavior that helps perpetuate the species. Historically, families with protective fathers produced more children who survived than families without male protectors. The biological genes of the protectors, therefore, along with the protection pattern, were more likely to be expressed in the next and succeeding generations. If men want to increase their progeny's potential and ensure their gene pool's continuing existence, it is in their interest to remain in an adult pair-bond; it is the surest way to strengthen their children's chances for survival and success.

But there is a caveat. Given the needs of the workforce and the needs of modern families in which both parents work, we are unlikely ever to revert to the days when the little woman stayed at home polishing the furniture and her husband went to work to earn the bacon. Nor is a mother likely to threaten her misbehaving children with "Wait till your father comes home!" or greet her conquering hero at the door with a kiss and a martini.

These have always been unrealistic scenarios, and besides, both husband and wife arrive home from work at about the same time.

The point is that if both parents must work, a successful family needs an equitable distribution of labor at home. Study after study has shown that women, fully as tired as their husbands after a day at the office, continue to be responsible for the major portion of housework and child care.

Gone are the days when a caring, protective husband could justifiably limit his work at home to mowing the lawn on weekends, cleaning out the garage occasionally, and teaching his son to play catch. And good riddance. "Woman's work" *is* never done, and his wife needs his help. Changing diapers and doing the laundry will not undermine a husband's masculinity; in fact, taking on household chores willingly and doing them well adds to his attractiveness in his wife's eyes. He will even know where to find his own socks, and that, according to a woman of my acquaintance, makes him a priceless gem.

Many new demands and stresses burden today's families, so it follows logically that self-interested young men shy away from marriage. But men who don't, those who can take the long view as well as the short, understand its value and understand that it is good for them.

Their own families' patterns tell them so, their
genetic patterns tell them so, and the women they love
confirm this truth.

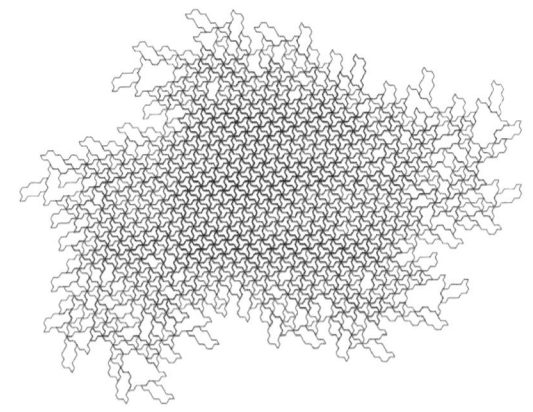

PREJUDICE
AND DIMENSIONALITY

AS WE MATURE, we are forced to confront prejudice, a deeply implanted pattern that appears to be universal in humankind and that does much to keep society divided. The roots of prejudice are found in early childhood, when children begin to categorize and classify in order to make sense of the huge number of stimuli that otherwise would overwhelm them.

Little children classify simply: people are adults or children, boys or girls, family or non-family. Dogs and strangers are friendly or frightening. In adolescents, the brain continues its pattern of classifying so that young people can create order from the chaos of constant stimuli, more and more of which need to be sorted and filed into more and more classes and categories.

Since civilizations began, people have had a surprisingly accurate grasp of this phenomenon. Medieval Europeans, for example, lacking books or file cabinets (or for that matter anything to file in them) but needing to learn and remember, thought of their minds as houses with many rooms, each room devoted to a particular category. When a sub-category or a new category required additional space, another room was added to the house. It was a pretty efficient system for the time. Today, lest we humans founder in an ocean of information, computers have taken over much of the job.

There is a shortcoming, however, to the traits we use for classifying and indexing: they are only a portion of the classified item's characteristics. (If you ever have tried to index a book, or tried to find something in an illogically assembled index, you understand this problem.) That is, although classifying focuses on certain characteristics, it ignores others, and these others may be the very ones that vastly distinguish one item from another.

Because relying on an index is easy and thinking our way past an indexing failure is hard, we may respond only to the indexed characteristics and treat diverse items, including people, in identical ways. Prejudice is essentially a result of improper, inadequate indexing.

Take for example a friend of mine who grew up in the segregated South. She is bright, highly educated, and liberal of outlook; nevertheless, as she has acknowledged ruefully to me, whenever images of black lawbreakers appear on the TV news, the old indexing patterns kick in and she must try *consciously* to relegate them to the trash heap where she knows they belong. It is an almost physical effort, she says.

Old, noxious patterns never die entirely; when you least expect it, they rear their ugly heads. Analogously, when you deposit those no-longer-useful documents and files into your computer's trash can, you like to think they are gone forever. But your computer, like your mind, doesn't really destroy the trash; it's still there, clogging up your nice, logical hard drive, and it can be retrieved.

Human beings seem unable to avoid mistakes in indexing. We are by nature ethnocentric. The best way to offset the inadequacy of our cataloguing system is to recognize our own prejudices and bring them forward into the cortical level, where in the bright light of our intellect we can examine them and determine whether they are improper, unfair, or unkind to the object of prejudice. The difficulty in changing patterns of prejudice is that they are embedded in the subconscious and acted upon in response to those misclassified stimuli. We see slanted eyes or dark skin or a fat body and immediately feel antipathy toward the whole person, ignoring everything else that makes that person someone whose company we might enjoy.

Societies, too, hold deep-seated prejudices about one another. In this regard, recall what happened after the collapse of the Union of Soviet Socialist Republics. For seventy years the totalitarian regime had tried to sever racial and tribal ties by moving millions of people to different parts of the Union.

They were transported, indoctrinated, forbidden to practice their religions and ethnic customs. Yet with the break-up of the USSR, old patterns of racial, ethnic, and religious conflicts reappeared—intact, and strong as ever. The pattern of prejudice was so dominant that it flourished by itself in the minds and hearts of people even when they themselves were separated. Once a pattern has taken root in a society, it is nearly impossible to exterminate, for the virus is passed to each succeeding generation.

Prejudice is abetted by a concept I call dimensionality. It measures the depth of our emotional relationship with others. Third-dimensionality, the highest level, involves strong emotion and develops with close contact. Third-dimensional people are usually our closest family and maybe one or two very intimate friends. They also may be our bitterest enemies, the people we hate. Third-dimensional societies include our country, state, and town, our church, and intimate clubs and groups; these are bodies with whom we identify.

Second-dimensional people, on the other hand, may be close and important to us, but the emotional ties are less strong. Among this group are friends who move in and out of our lives, people we know from the office or from school, people on sports teams of which we are members; they share our interests and we like their company. Both individuals and societies are subject to dimensionality. Americans, for example, view the United States and other Americans as third-dimensional. The French see their country and their countrymen in the same way.

Second-dimensional societies, on the other hand, are those which with we interact on an emotional level and care about or dislike with some intensity. The United States, for reasons of intertwined history, religion, and politics, is oriented primarily toward Europe, particularly

Western Europe. Israel, an essentially Western nation despite its location in the Middle East, is for many of us a second-dimensional society. Some Americans, especially those who can remember the Second World War, continue to feel considerable rancor toward Germany. Interestingly, I think, among European nations England is an exception, as most Americans see England in three dimensions. Our nation's history is inextricable from England's.

We fought each other in the American Revolution, then fought together in two world wars. Our political systems are very similar, we have allies and enemies in common, and perhaps most important, we speak the same language. I believe, in fact, that our third-dimensional view of England may be the basis of our "special relationship"— a relationship that we all sense as real but that no one seems able precisely to define.

In our personal lives, we see fewer dimensions as we move farther outside our familiar circle, and we treat the rest of the world first-dimensionally. This world is like a movie, and we are the spectators. Generally, we don't associate with first-dimensional people. We don't identify with them or invest emotionally in them. At most, we may feel twinges of pleasure or dismay in watching something very nice or very unpleasant happening in their world.

Probably the majority of Americans perceive Africa and Asia in one dimension, so that even as television brings their problems into our living rooms in vivid color, we tend still to be only spectators. We see African children starving, Palestinian youths buried alive, Israeli children blown to bits, whole populations obliterated in a tsunami. Television shows us groups without number or name. They are first-dimensional people; we know they're there and we're briefly sorry they have such problems, but we feel little prolonged concern about them. Does this response mean that we humans are a cold-hearted race? Not necessarily. Dimensionality is useful as a protective device. If we could feel every person's pain, we couldn't function; the burden would be too much to bear.

Yet our dimensional view of the world may be changing. Some believe that our constant exposure to the tragedy and horror that dominate television news desensitizes us. I think, though, that in time it may produce a cumulatively good effect. Moreover, in addition to television, social media such as Facebook and Twitter give us the immediacy of direct, person-to-person contact. And after all, most of us have received some patterning that includes the concept of caring for others; as we observe world conditions, we may change the dimensionality of our perceptions.

It is a slow process—slower in some people than in others—but once a majority of the society accepts a new dimension, sees it as the norm, then we can effect social change. Obviously, as in much re-patterning, such change requires emotional stimulus.

For example, television recorded galvanizing moments during the Civil Rights movement. We saw an Alabama sheriff urging his dog to savage black people peacefully participating in a march. We saw women and children mown down with fire hoses. We learned of the murder in Mississippi of four white voting rights workers. These and similar images lifted the concerns of blacks in a white society to the forefront of our minds and moved many people from first- to third-dimensional consciousness.

When terrorists struck the Twin Towers, all Americans shared the shock and vulnerability of the people of New York—some nine-million souls who, a moment before the planes struck, had been first-dimensional strangers. Quite recently we watched as millions of Egyptian citizens staged an eighteen-day peaceful demonstration that ended President Hosni Mubarak's thirty-year dictatorship. What American with even a rudimentary knowledge of his own country's history could fail to identify with those revolutionaries in Egypt?

Dimensionality can shift—slowly, as in the years of struggle that led to civil rights for blacks in America and that culminated in the election of President Obama, and very quickly, as in the Egyptian revolution. When we enter into war, the enemy may begin as a first-dimensional entity, but when hostilities start, the enemy moves into the third-dimensional realm. We focus as fervently on our enemies as we do on our loved ones, and during a time of war, a society to which we have given minimal attention becomes hugely significant to us. North Vietnam? Where's that? We soon found out.

Sometimes a mesmerizing leader can move a population to consider dimensionality as an issue. President Kennedy's creation of the Peace Corps, for example, changed forever the perceptions of those who participated. Americans found themselves in strange places among strange people with strange customs, and never again could Peace Corps volunteers see those people and places in one dimension, for they had become "real."

That pattern continues today in new Peace Corps volunteers, and as they share their stories, they change the perceptions of those they encounter at home. It is the ripple-in-a-pond effect. If there were enough ripples in enough ponds, if we as a society could come to view all human beings in three dimensions, then the force of the tide would be overwhelming.

We would see massive pattern changing, and perhaps—dare we dream?—a more peaceful world.

THE PATTERN OF
HUMAN AGING

BROADLY SPEAKING, aging is a species-specific activity. That is, an insect may live only a few hours and a whale many decades. But each species' lifespan appears to be part of its genetic inheritance. Scientists who study aging have yet to ascertain exactly what causes aging, although numerous theories have been advanced. Some researchers think that aging occurs because of errors in the translation of DNA to protein: as an organism ages, the errors increase and the body cannot repair itself efficiently. Therefore, organisms with the healthiest genes and strongest repair systems live longer than those that don't.

Other theories involve telomeres, which are repeated DNA sequences, or cellular "caps," that protect the ends of chromosomes and control how many times cells can divide. And as they divide, their telomeres shorten; when they become too short, division ceases and the cells die. It used to be generally accepted that, theoretically, cells were immortal, that they could divide endlessly.

In the early Sixties, however, University of California researcher Dr. Leonard Hayflick learned through his work with lung tissue that indeed there is a limit to cell division: fifty or so divisions and they died, and before they died they aged. Many biologists now believe that aging results from telomeres' shortening as more and more cells reach their "Hayflick limit."

Theorizing about the causes of aging is interesting, but as yet it is of little practical value to those of us who confront the realities of growing old. How shall we deal with it? The advertising industry offers a wealth of suggestions: hair restorers, wrinkle creams, age-spot removers, vitamin concoctions, "colon balancers"—the list is endless.

Pharmaceutical companies implore us to try their latest cures for every disease in existence and for some that aren't. Given the monumental number of pills and potions and palliatives on the market today, you'd think we could all stay forever young. And yet, quite probably before we are born, the pattern is set; inexorably, we age. But although currently we may not be able to slow the aging process without encountering some serious unintended consequences (it is alarming from a medical point of view, for instance, as well as just sort of unseemly, to imagine an eighty-five year-old Viagra-popping man sporting a four-hour erection), there are ways to make it more palatable.

The first imperative is to continue to set new goals. People need thrust to their lives. It doesn't matter what the thrust is as long as it matters to the person and demands a pattern of repetitive action. Attractive as retirement may seem to someone who has worked hard all his life, it is a destructive concept, because if his idea of retirement is merely to "relax" and "have fun," he will deteriorate quickly and visibly.

It seems healthy people who have quit one job and moved on to another passion, whether it is paid or unpaid work, usually maintain their alertness, their interests, and their brainpower. They age less perceptibly, maybe growing a little grayer and a little more arthritic, and they remain vibrantly human.

I'm acquainted, for example, with a Franciscan nun who retired several years ago after thirty-five years as a teacher.

Shortly after settling in the small, backwater town of Grants, in remote northwest New Mexico, Sister Martha observed that the local animal shelter was a disgrace—a broken-down affair of rusty pens and sheet metal and chicken wire and mud and vermin, hidden out of sight and out of mind on the ragged edge of the city. The place posed health hazards both to its employees and to the animals it supposedly sheltered.

Sister Martha put her intellect, her considerable persuasive power, her energy, and her love for animals to work. She reestablished the town's defunct animal welfare group. She visited possible sites for a new shelter and dickered with their owners about cost. She researched funds-granting agencies, wrote a successful a proposal, then persuaded an obstinate city council to provide matching funds and request bids from architects and contractors.

The result?

Although the entire process took five years of Sister Martha's tact and cajoling and persuading and plain, persistent hard work, it all paid off. Today, situated on a prominent, easily accessed site stands an attractive, state-of-the-art animal shelter that is a source of pride to the community and the envy of other, larger New Mexico cities. And I'm sure that during her intrepid quest for a new animal shelter Sister Martha confronted challenges and obstacles new to her, and that she had to develop new patterns in order to overcome them. After all, an unassuming nun who has spent her life teaching arithmetic to children isn't necessarily prepared to take on contractors and a small-town oligarchy and seven thousand dubious citizens.

Death is everyone's destiny, but the art of living is to perform to the full extent of our abilities until death arrives.

As the debilities and failures of our physical selves progress, we have to address the issues that these failures generate, but they needn't destroy our motivation. The key to a good life is to keep moving toward goals, and the patterns of work that we established long ago will help us achieve them. Work itself creates endorphins, which in turn permeate our being, and like a good narcotic, ameliorate many of our symptoms; with something more important to think about, we forget about them.

New goals, moreover, even may require us to establish new patterns, a salutary outcome almost guaranteed to enhance and possibly to prolong life. Colonel Sanders, for example, started the immensely successful Kentucky Fried Chicken franchise when he was in his sixties, and at about the same age Lillian Carter, President Jimmy Carter's mother, joined the Peace Corps. In her seventies, arthritis forced Grandma Moses to abandon her embroidery career; she took up painting instead and became, well, Grandma Moses.

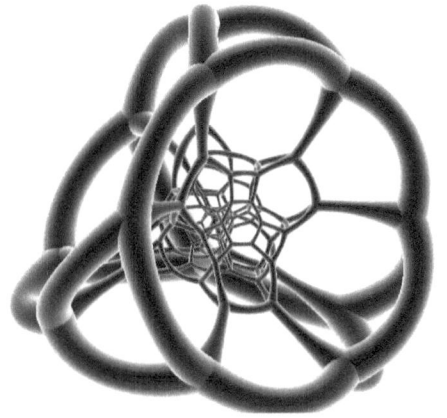

CAGES

JADE IS THE most difficult of gemstones to work with. Seen under a microscope, jade is a mass of matted, intricately interwoven fibers that, taken together, produce a tough stone nearly impervious to fracture. In fact, what we think of as jade "carving" isn't really carving at all; more accurately, it's a wearing away of the stone so slowly that a design emerges.

In ancient times, it was immensely time-consuming. Today's jade carvers use modern implements and technology, and their work takes less time. The "wearing away" principle still obtains, however, because the nature of jade itself has not changed.

Patterns are perhaps the most difficult element of the human condition to amend. Not only are genetic and perceptual patterns intricately interwoven—much more so, even, than jade fibers—but they acquire a life of their own, helping us respond efficiently to the torrents of stimuli that we encounter every day.

It requires extreme external pressure and a great investment of emotional energy to modify even the simplest patterns. Just ask any psychologist. Despite new, effective, and less time-consuming psychotherapies as well as pharmacological methods of helping people alter destructive patterns, the nature of patterns is to resist change.

And making pattern change even harder is the fact that patterns also are enculturated. That is, each of us lives in a succession of "cages." The word isn't necessarily pejorative. Cages are the sum total of the patterns and perceptions of a person or a group.

For example, there is the cage of one: the individual and the patterns and perceptions that are exclusively theirs.

But because people aren't hermits, they live in other cages too, such as the cage of family, of community, of tribe, of religion, of race, of nation. I begin as I, but as I mature I inhabit more cages. I'm a son, a husband, a father, a Jew, an Ohioan, an American, a physician, a businessman. I inhabit cage after cage, some in succession, sometimes simultaneously.

These cages are like Russian nesting dolls, with each one successively larger to accommodate and enclose all the others. For societal change to happen, light and information must penetrate the many cages. Further to complicate things, the cage walls, which comprise the patterns and perceptions of all who live in the cage, are rigid, semipermeable, and semi-reflective. Residents of the cages spend most of their time contemplating old perceptions, not considering new ones. They look inward toward what they know, rather than outward toward what is new. Too often, a new insight "reaches the heart, and dies."

Societies comprise multiple interconnected cages. Each of the cages in which we live has locked down its perceptual patterns, and each group in its cage functions primarily in response to the stimuli it allows into the cage. Moreover, the actions of those who reside in the cages are those permitted by the others who share the cage and by the boundaries of the cage itself. Only a limited number of stimuli will penetrate all of a society's cages, and some cages will not perceive most stimuli at all.

Here is an example.

Warren Jeffs, the self-anointed "prophet" of the polygamous Fundamentalist Church of Jesus Christ of Latter-Day Saints (the FLDS "church"), was convicted of child rape; a Texas court sentenced him to life plus twenty years. The rape victims were twelve and fourteen at the time of their "spiritual marriages" to Jeffs.

DNA from the older girl's baby proved Jeffs was its father. In addition, jurors were treated to a "wedding" photo of Jeffs kissing his twelve-year-old "spiritual wife" as well as a tape recording of Jeffs introducing his little "bride" to the pleasures of sexual intercourse with a fifty-five-year-old man, namely himself. It is not known exactly how many "wives" Jeffs has acquired during his years as FLDS leader. Some say as many as sixty; others say there are more. In this sect, apparently nobody's counting.

Meanwhile, back at the FLDS compound, what must Jeffs's FLDS followers be thinking? With their leader convicted of heinous acts and in prison for life, might it not be appropriate to reconsider their principles, not to mention their relationship with Jeffs and with one another?

Not at all.

For FLDS members, indoctrination begins at birth. Their schooling is minimal, and they are locked away from the outside world's "corruption." (Jeffs forbade television and Internet use while his trial was taking place. They obeyed, and discounted as false any news that may have filtered in.)

Although a few courageous women, and even fewer men, have escaped the sect and are trying to shine a light into the darkness they left behind, almost all FLDS members remain in the cage. With renewed strength drawn from their leader's perceived martyrdom, they continue to believe what they have always believed—and *only* what they have always believed. After the Jeffs trial ended, news programs reported that his followers plan to erect a statue in his honor—a thirty-eight-foot statue of Jeffs holding a little girl's hand as she looks up at him. Trustingly.

If there is sufficient energy, cages can be blasted apart. When the doors open, the inhabitants must move fast toward change before the walls recreate themselves. If change doesn't take place and time passes, the cage is reconstituted, often to be even stronger than before.

The Supreme Court's 1954 *Brown v. Board of Education* decision, for instance, mandated racial integration of public schools. The decision angered a great many Americans, by no means all of them Southerners. My friend from Maryland, whom I've mentioned before, sent me an e-mail describing what was for her the most memorable incident of that era.

I didn't understand most of this thing that happened until years later.

I was spending the summer of 1954 with my grandparents. I was twelve. For a long time, probably a year or more, there'd been talk about the *Brown* case—everybody knew what the decision was going to be—and what the county schools would do about it. Some people swore they'd never send their children to an integrated school. A few people even mentioned secession as a possibility, but they weren't really taken seriously. And it wasn't just white people who were jittery about integration—so were the black people. Everybody was expecting trouble.

You have to remember, this was a time when doctors' offices and train stations had white waiting rooms and "colored" waiting rooms. ("Colored" was the polite term then.) There were white and colored public bathrooms and drinking fountains, white and colored schools. The whole thing sounds like ancient history now, doesn't it? But it wasn't that long ago.

My grandfather was chairman of the county school board. A colored man named Harris was on the board too, representing the colored schools. The board members were pretty sensible men, doing their best to plan for the social upheaval about to begin. Granddad's most persuasive conversational ploy, I remember, was to point out to black *and* white, "You know, we don't have to like it, but if it's the law of the land we have to make the best of it." I would hear people repeat that to one another, as if Moses had spoken. But mostly I was oblivious of the political hurricane pounding the country. I'd discovered literature, so I was too busy reading.

Anyhow, on a summer morning shortly after the *Brown* decision came down, I was stretched out on the front porch swing, reading, when my grandfather interrupted me. He said he was expecting a visitor, a colored man named Mr. Harris, and I was to run inside and let him know the minute Mr. Harris drove up.

In a brown car.

Let him know right away.

"Mr." Harris? Nobody called "colored" men "Mr." While I was puzzling over this unusual request, I heard voices raised inside the house. Well, one voice—my grandmother's. Nana was shouting at Granddad. She *never* shouted at him. Her preferred method of getting her way, like most women's then, was passive aggression. I heard her stomping through the house. She flung open the front screen door and glared at me. Her throat was a mottled red, and her whole face—chin, cheeks, and forehead—was blazing pink. She turned back inside, slamming the door behind her. Nobody in that house ever slammed doors. This was getting interesting. I listened harder.

Nana: "Not at my dining room table! I won't have it!"

Granddad: "Lalie, it'll just be a cup of coffee."

Nana: "He can have a cup of coffee in the kitchen."

Granddad: "Lalie, I invited him. I *can't* ask the man to sit in the kitchen."

Nana: "If he's so important, you can sit there with him then."

Granddad: "Lalie, be reasonable. *Please.*"

(My grandfather never pleaded. Not for anything.)

Nana: "Reasonable? Reasonable? Well here's *my* reasonable. *No damn nigger will EVER* come in my front gate and walk in my front door and sit himself down at my dining room table! And damn you too, Sam Townshend, if you try to let him do it."

(Wow. Our family didn't curse. And nobody was allowed to say "nigger," either. Nana's edict? It was everybody's edict and everybody obeyed it, but nobody ever said it out loud.)

Just then a colored man in a brown car drove up and pulled in close to our fence, parking about equidistant from the front and back gates. He sat there, not opening the door. Before I could hurry and tell Granddad that Mr. Harris was here, the back door banged open and I saw my grandfather *running* (he never ran), on our side of the fence, to greet him. Mr. Harris got out of the car, they shook hands, I heard Granddad saying apologetically that Mrs. Townshend had a bad headache, so they'd better have their discussion here, outside. Mr. Harris said he understood. You bet he did.

They talked a long time, standing there, but I don't remember anything more of what they said. Probably they were confirming final plans to begin school integration. What I do remember is the two men, black and white, separated by so much more than the fence between them, and how earnestly they were trying together to make the best of a terrible situation.

THE CAGE OF HISTORY

WHEN "things have always been this way," people are naturally reluctant to change them. If you're reasonably comfortable with the life you're living, but you see major change looming, logic and common sense can give way to emotion; you may reject the very thought of change no matter how much benefit it may bring.

It is doubtful, for instance, that either my friend's grandfather or Mr. Harris would have sought school integration without the impetus of *Brown*; faced with it, however, they stepped outside the cage of history and cooperated to accomplish a result that would serve their community's best interests.

The grandmother was a different story. "Well, Nana was from Virginia," my friend says in her grandmother's defense, "and Virginia women were different. Besides, she was an FFV." ("She was a what?" I said, wildly considering possibilities.)

"An FFV," my friend said. "You know—First Families of Virginia."

Oh.

Nana's patterns of thought were set in granite, and to challenge them would be to challenge her sense of herself. Locked in the cage of history, Nana *could not* allow a black man at her dining room table; it was not possible; it would shatter her identity.

Much as we may deplore the thinking of the Nanas in this world, these poor caged creatures also deserve our compassion. What if Granddad somehow had managed to force the issue and despite his wife's objections had invited Mr. Harris into the dining room for coffee?

I'm certain that Granddad knew it would be a Pyrrhic victory, won only at terrible cost to the marriage.

THE MARRIAGE CAGE

"I AM I because my little dog knows me," Gertrude Stein famously said. People need someone else in their lives to reflect and confirm their identity. If another person isn't available, a dog, as countless humans have discovered, will serve quite well—maybe too well, as your dog is likely to confirm everything good that you believe about yourself and forgive all your faults, thereby creating a more flattering image of you than you perhaps deserve.

Absent a dog, maybe a cat will do.

In my experience, those opaque cat-eyes don't reflect much of anything; if you are a cat lover, though, you probably think they do. But what about a volleyball?

You may have seen the movie *Cast Away*. Tom Hanks plays the protagonist Chuck Noland, isolated for four years on a deserted island, his only companion a volleyball. Early in the story Noland accidentally cuts himself, and blood splatters onto the ball; Noland fingerpaints a vaguely human face into the blood and names the ball Wilson. As the years pass, Wilson takes on increasingly human characteristics. The entire metamorphosis occurs in Noland's mind, of course; nevertheless, a quite genuine man-volleyball relationship develops.

Wilson becomes Noland's friend, a "person" with whom to share hopes and fears, to ask for advice and opinions, to "come home to" after an excursion around the island.

Noland argues with Wilson too, sometimes angrily, once even going so far as to kick the ball out of their shelter—but then immediately retrieving it. Nothing seems demented about their relationship; Noland is perfectly aware that Wilson is just a volleyball, yet a very *important* volleyball. In Noland's desolate state, Wilson keeps him grounded.

Consider the marriage cage: two people bound together by law and custom, by love, by the children they produce, and often, like Nana and Granddad, by many years of life together.

They come to know each other in countless large and small, spoken and unspoken ways. For a marriage to endure successfully, the relationship requires balance. Inevitably there will be some disagreement, some friction. Both husband and wife must be willing to relinquish part of their autonomy, yet by the same token each needs to recognize and respect the other's limits. Sometimes the husband's preferences will win the day, sometimes the wife's.

When it is a question, for example, of watching *Sunday Night Football* instead of *Masterpiece Theater*, or buying a Honda instead of a Ford, a married couple usually either compromise, or one of them gives in without too much fuss. But when it is a matter of the essential self, of who one believes oneself to be, each understands that the other has boundaries that cannot be breached without seriously endangering the relationship.

If a Catholic marries a Muslim, for instance, neither of them scorns or trivializes the other's religious beliefs, as they are so profoundly basic to identity. To Nana, an entirely unreconstructed Virginian, the very thought of a black man enjoying coffee at her dining table was anathema; it would have turned her world upside down; it would have told her that after all these years of marriage, Granddad had no respect for her feelings or their home.

Worst of all, it would have destroyed her sense of self. The fact that her identity was constructed on very shaky ground in the first place made it all the easier to bring down. At a deep, intramarital cage level Granddad must have understood all of this, thus his choice of discretion over valor.

THE CAGE OF SELF

THE FIRST CAGE we inhabit is the cage of self, and that cage is unique to each of us. The trees you see are different from the trees everyone else sees. The grass is a different green. There is a world outside the cage, but the world the individual person perceives comprises those things that pass his perceptual filters and are interpreted through his senses. Each person creates a cage of self, and often the walls of the cage seem to be the walls of the entire world.

A newborn baby arrives in the world with a cage already in place. The embryo has developed in a tight genetic cage donated by its parents, and that embryo maintains its genetic inheritance; a frog never becomes a wolf. In almost all mammals, the embryo grows in its mother's uterus, which is ordinarily a safe and secure cage.

At birth, the child escapes that cage only to enter another—the cage that its parents immediately begin to construct for him by teaching him how the family cage operates. The child himself becomes involved in cage-building, using his senses: feelings of fear and pleasure, hunger and satiety, abandonment or safety, rejection or acceptance, stress or contentment—all these things reinforce the bars of his cage.

During the first three years of life, the child sequentially bursts through barriers: from crying to laughing, from babbling to speaking, from lying to sitting to crawling to walking. Although he remains within his primal cage, he stretches its boundaries, forcing the cage to grow with him as he moves through the stages of development. Jean Piaget, an early leader in the field of early childhood development, has suggested that small children are like little scientists who construct and then test out their own theories about the world.

That sounds about right.

By the age of three, a child has put together a cage of experiential boundaries. Within it, he has learned to give and accept love—or indifference. He has learned to fear and to frighten. He has a rudimentary comprehension of family and social cages that has been transmitted to him by his parents and other significant persons. As the child grows, his understanding increases, but that early-formed cage is always with him, a structure that includes biases, beliefs, and attitudes that will endure the rest of his life. That is why, at thirty, fifty, or eighty, you so often catch yourself "acting just like" your mother or father.

In order to maximize a child's intellectual and social growth, parents should offer their child the broadest possible spectrum of experience so as not to restrict his intellectual and social growth. The more—and the more varied—his learning, the wider and more permeable the walls of his cage, and the better equipped he will be for life. If two children have similar genetic endowment, but Elaine lives in a stimulating environment and Anna grows up in deprived circumstances, Elaine will not become "smarter" than Anna, but she most likely will be socially and intellectually nimbler.

An acquaintance who spent ten years as an instructor at a small Navajo community college tells me, for example, about the differences he observed between students who had spent their entire lives on the reservation and those who had lived in distant cities.

"The reservation is an enormous, remote area," he says, "with long distances between the few little towns, and real cities are all far away from the reservation. There are no movie theaters or even any restaurants except for two or three at tourist sites, and the very occasional McDonald's.

Huge numbers of people are jobless and poor. They don't have telephones or indoor plumbing or much hope of a future. My students from the rez were bright enough, but they—I don't know—they seemed intellectually stiff, somehow. They had a lot of trouble understanding new ideas. It was like they just couldn't flex their minds easily. Whereas the kids who had lived elsewhere usually caught on to things much faster and made better grades. Socially, they were the leaders."

The teacher noticed similar differences even between many Navajo students and those from the Pueblo tribes. "For one thing," he says, "most of the Pueblo Indians live closer to Albuquerque and Santa Fe. They work in those cities, so they rub shoulders with all kinds of people every day. But the most important difference, I think, is in the Pueblo culture itself. They have hundreds of years of experience in living close together—"pueblo" means "town," you know—and hearing everybody's ideas and opinions, and cooperating to get things done."

Only a fraction of what goes on inside most people's cage ever leaves it, and only a fraction of what happens outside ever gets in. Sometimes an aperture opens, and a person gets an unobstructed view of the societal cage in which he lives; more rarely he catches a glimpse of the world beyond.

Maybe the ideas in a book penetrate his consciousness, or he develops a friendship with someone from another culture. And travel really *can* be broadening. I've known military families, for instance, who after years

259

of postings from country to country come home bearing a wealth of new perceptual patterns regarding the world.

These are thoughtful, interesting people, agile of intellect and fascinating to converse with. Unfortunately (they tell me), the majority of military families abroad choose to cluster in "American ghettoes" rather than live among the "foreigners," effectively blocking the new and the different from entering their cages.

Except for a hardened belief that all things American are best, they return home unchanged. What wasted opportunities, and how typical of so many people.

The carapace of self both protects and restricts the creature within. This persistent dualism results in a continual tension that informs every aspect of our lives. Does the quarterback risk a Hail Mary pass that can win the game, or are the chances too great that he will be sacked and possibly injured by a 340-pound linebacker? Does the politician tell the truth, or, fearing a humiliating loss, does he fudge the facts?

Do you dare ask Sylvia to marry you, or does dread of rejection prevent you?

Always, we are weighing the possibilities to determine just how far outside our safety zone to venture without endangering our selves, and each of us calculates the odds differently in different situations and at different times. Most of us, most of the time, are timid souls, yet each of us is capable of the occasional Hail Mary.

In order for society to function, we have to enter the cages of others, toting our own cages along with us. The result is a sort of living Venn diagram—those intersecting, overlapping circles that show relationships among sets of things. Once there, we perceive only what is within the overlapping portions of our cages; we are blind to the parts outside the overlap. The intersection of cages allows new ideas but closes off others until we move into another cage in which they appear again (theoretically the number of intersecting circles is limitless).

The movement of the self from cage to cage affects our behavior. We adopt various personas to fit in. At work, for example, a woman is a high-powered advertising executive, answerable only to a higher-powered executive. At the PTA meeting, she presents herself as a concerned mother. At the garage where she is leaving her car, there may be very little cage overlap; unable to grasp most of what the mechanic is telling her, she becomes a supplicant: "Please, Ed, just try to fix it. And do you think I can have it back by Wednesday? Please?"

At work, this woman's current project is to create a campaign that entices young people to take up smoking. Her activities reflect the realities and mores of her work cage: she is providing a service for a client, earning a living, and helping to support a family. When the workday is over, however, she may take her place on a hospital board that discourages smoking because of its effects on people and on health-care costs. She heads a committee that designs antismoking strategies. She is an effective board member, advancing the interests of the hospital, and she is seen as a leader in this arena.

Even though her community cage and her work cage are in conflict, she nonetheless can operate in both worlds by adopting the perceptions of the cage-of-the-moment; that is, she compartmentalizes. The woman is trying hard to deal with cognitive dissonance, that psychologically discomfiting circumstance wherein you attempt to hold two opposing beliefs at the same time. If the dissonance becomes too great, she will have to make a change in order to live with herself; she will resign the account or leave her board position, thereby regaining homeostasis.

THE ADDICTION CAGE

WHAT IS ADDICTION? A sign of weak character? A habit that anybody with an ounce of will-power can break?

A disease? Why, for instance, is John satisfied with two beers, but Jim has to down the whole twelve-pack, or why can Martha leave the slot machines after losing fifty dollars while Linda hocks her jewelry to try to "win back" her losses.

We don't know, precisely, the causes of addiction, and our treatment methods have a dismal failure rate. Some people reject the disease model of addiction, believing the addict is simply a weak, inferior person—an unhelpful attitude that merely confirms most addicts' low opinion of themselves. Nancy Reagan's Just Say No anti-drug campaign epitomized the simplistic will-power approach and are among the most intractable of pathologies and the most impermeable of cages.

Today, most physicians and other professionals in the field accept that drug or alcohol addiction is a chronic, progressive, and in many cases fatal disease. An armamentarium of psychotherapies and pharmacological therapies are available to treat the addict, as well as self-help groups such as Alcoholics Anonymous, but results are mixed at best.

What makes a person vulnerable to drug or alcohol addiction? I believe the cause may lie in the primal cage, which is not always inviolate or completely successful. There may be a genetic flaw, a deficiency of attention or stimulation, a lack of love. External circumstances such as starvation or disease may play a part. Incest and harsh physical abuse are major precursors of addiction. When the cage comes under assault from any of these pressures, it may buckle, leaving the child far more vulnerable to addiction.

The addictive substance penetrates the cage of self, piercing the addict's very nucleus and becoming part of his identity. And although addicts often are dealing with issues that arose during the formation of the first cage, addiction is by no means limited to those who endured childhood trauma. Millions of American veterans suffering from post-traumatic stress disorder have become hopelessly addicted to drugs and alcohol. Many of them today may be homeless—the most marginalized of our citizens.

The addiction cage is one of the few that a person carries into all other cages. The addiction affects all his interactions because it has skewed his sense of what is appropriate, so much so that he may be comfortable only in the cage he shares with other addicts.

In fact, so powerful is addiction that a person essentially stops maturing when the addiction takes hold. If after years of addiction to alcohol he somehow summons the strength to stop drinking, he may find himself ill at ease in a world that has changed since he first started.

I have a friend who began drinking at twelve. At forty-seven, after several prolonged stays at rehabilitation centers, untold hours of therapy, and innumerable AA meetings, she finally stopped.

But whereas in her drinking days Julia was always the life of the party, today, even after several years of sobriety, she avoids social gatherings. "It's not that I'm afraid I'll start drinking again," she says. "I'm confident that won't happen. It's because I feel like a kid at a party for grown-ups. I'm afraid I'll do or say something stupid and then everybody will laugh at me."

There are other addictions besides drugs and alcohol, some of them more socially acceptable, but all of them harmful. Many of us are addicted to gambling, to food, to smoking, to television, or to the Internet.

The gambling addict, for instance, may lose more than money; he may lose his home and his family. The food addict destroys his health. The smoker is vulnerable to a host of terrible diseases, many resulting in death. The TV addict isolates himself from family and friends so that he can enjoy the false intimacy of television; his "real" life is happening on the screen—but only in two dimensions.

Inhabiting the ever-expanding universe of cyberspace is a relative newcomer to the field: the Internet addict. And who can blame him? In addition to its value as a communication (to everyone! everywhere!) medium and as a seemingly limitless source of information, the Internet offers virtual love to the loveless and virtual power to the powerless—and all this at a keystroke. Like other addictions, the Net is extremely seductive to susceptible people.

People often say that an addict "is hurting only himself." Would that it were true. But the fact is that the addict hurts others as well. His family suffers; his work relationships suffer. His addiction-incurred health problems add tremendously to the costs of medical care for us all. The addict, his family and friends, his doctor and his psychologist all may be giving his condition their best efforts, yet the disease frustrates them. Without knowing the cause, it is hard to find a cure. Advice to "be moderate in all things" is pointless, as addiction itself defines immoderacy. Waiting for him to "hit bottom" before effective treatment can begin seems monstrous; would you tell a woman with stage 1 breast cancer to wait until stage 4 to consult an oncologist? At present, apparently the only feasible choice is to keep doing what we're doing, hoping for good outcomes but expecting setbacks, and dealing with them as they arise.

THE FAMILY CAGE

SIMILARLY PATTERNED ORGANISMS build cages, the most intimate of which is the family cage. The family maintains its own perspectives, attitudes, beliefs, and practices, and brings intense pressure to bear on its members. This pressure is so great that it precludes any but a few from ever escaping the cage. If indeed a person does break from the family, he may find himself unmoored, awash in a sea of self-doubt. Because his flight was reactive and hasty rather than considered, he has left the family cage without having constructed a sufficiently strong new cage within which to feel safe.

Although today we are told that a "family" can be anything we say it is (including for instance two men or two women or an unmarried woman and her cat), in our culture a family still typically begins with marriage. Human beings are pairing animals, and beyond that, tribal animals. The gravitational pull of the pairing instinct and of tribal identification is very strong. Weddings are universal, cutting across societies. The gathering of the clans, the happiness, the excitement, the gifts and food and drink and dancing and singing—all these are common to weddings, whether on Park Avenue or in a Bedouin tent.

Marriage merges separate pathways; two become one. Except that they don't, really. For a marriage to work, both husband and wife have to accept that radical pattern changes must take place. It isn't easy, but forming a family demands such changes. The young man and the young woman give up the patterns of their single lives in preparation for a different kind of life in hopes it will be a fuller, more satisfying life.

It is nest-building time. Particularly in couples who have not lived together before marriage, early marriage features an explosion of sexual activity, and the excitement and freedom of married sex create new patterns of togetherness and sharing. Relinquishing many of the pursuits that occupied them before marriage and learning to share love with one another, the couple prepare themselves emotionally to share love with their children.

Because patterns change slowly, marriage alone cannot alter them. It does, however, add energy that accelerates change. Although sex is the most dramatic energizer, it is not the only one. Simply living together helps couples change their patterns. For example, a man and a woman bring different food preferences to their marriage.

Eventually, they modify their preferences until they arrive at the family menu. It's easy when they both come from similar families with similar tastes and culinary techniques. Or if one of them (like me, for instance) had a mother (like mine) who was a terrible cook, and the other (like my wife) is a splendid cook, the culinarily-deprived spouse will send up grateful hosannas and change his old patterns in a flash.

But as people grow up, they generally like to eat what their family of origin ate, and eating patterns are strong. What is the Mexican-American husband to do when his Boston-born wife proudly serves a New England boiled dinner? To her palate, this is *haute cuisine*. To his, accustomed to hot, spicy dishes, the stuff has all the charm of damp library paste. It is an impasse that only love, time, and massive amounts of tact can break. It is marriage.

Traditionally, marriage alters many other patterns, too, and establishes new ones. Beliefs and values are modified from each spouse's original experience, and the melded patterns are conveyed to the next generation as "our family" patterns. Some of today's families are less likely, however, to transmit multigenerational patterns, in large measure because of the birth control pill. For the first time in history, a woman can fully control her reproductive life. No longer must she mate with the physically strongest man she can find for protection of herself, which she may not need, or of children she may choose not to have.

I mentioned earlier that a man needs to have a sense of ownership of those things for which he feels a long-term commitment. I have discussed the issue with several intelligent women, who singly and in chorus inform me that mine is precisely the kind of Neanderthal thinking that for centuries has kept women imprisoned in harems, covered by burqas, denied access to voting booths, and stuck under glass ceilings.

They inform me further that *some men have evolved* without diminishing their manhood. To these women, whose opinion I value, "ownership" is a far more negatively loaded word than I intend it to be. It is just that a man cares primarily for what he is committed to: his wife, children, church, and community. He is sympathetic to other people's problems, but his primary concern is to "his own." This is a genetic and societal pattern inherited from the distant past, when the difference between a community's life or death was possession of food, shelter, weapons, wives and children—all integral to survival—by a group of like-minded people.

For millions of years, men were the hunters. Aggressiveness helped him find food. A healthy dose of paranoia protected him from ambush, and, not incidentally, made him suspicious of other men's possible designs on his wife. (And inspired him to devise the chastity belt, one of those intelligent women remarks acidly. Point taken.) Possessiveness made him cling to the things and people that were his; without that sense of ownership, the tribe might perish. The men who survived, therefore, shared tendencies toward aggression, paranoia, and possessiveness.

Those ancient patterns have not disappeared, but women's newfound independence has made them seem inappropriate to marriage. Men know this—or if they don't, women will soon set them straight. Many men channel the traditional patterns into other activities, such as sports, either playing or, much more often, watching. When a husband spends untold hours communing with his favorite teams on TV, his wife may wonder, "Why this obsession with sports, night after night after night?" Well, it's because there aren't any mastodons left to kill.

Women's independence is increasing, not only in western nations but, perceptibly if more slowly, in many other areas of the world. The trend will not reverse. The major problem I see with it is that in undermining men's sense of ownership, it may also weaken their sense of loyalty to their wives and families.

Somewhere deep in the psyches of modern men, the old patterns remain, but the new roles that women have

271

assumed can cause men to feel literally "dispossessed." Very probably, we are witnessing the end of the family as we know it. For both men and women, the comforts as well as the constraints of the traditional family cage are diminishing, and we have yet to devise a suitable new cage.

Yet we must do something to support the children who become casualties of these weakened, radically altered cages. We know, for example, of successful role reversals in which the wife is earner and the husband is homemaker. Such unions are relatively rare, however, their numbers unlikely ever to cause societal change. In the belief that what worked for our ancestors will work for us, we might return to the traditional cage, but that approach belies the reality that we aren't our ancestors and we need to deal with the new circumstances in which we find ourselves. A third option is a return to the tribal cage, in which all members of the group accept responsibility for all the children. Although this very ancient form of family functions well in some small, homogeneous, and less advanced societies, it is hard to imagine its success in our enormous patchwork of a social structure. Perhaps our only course is to wait and see how human nature handles the new realities.

There will be families; we just don't know exactly what kind.

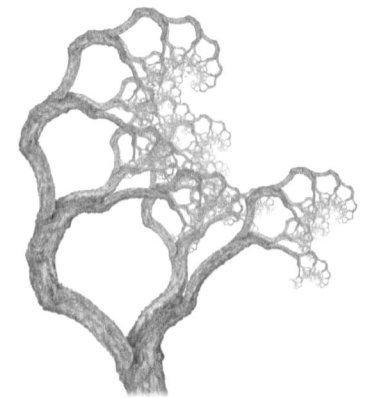

UNCAGING

I SOMETIMES REFLECT on the unhappy union of England's Prince Charles and Diana Spencer. At some point during their engagement, a reporter asked the couple if they were in love.

"Of course," simpered nineteen-year-old Diana.

"Whatever 'in love' means," muttered the Prince.

Love and romance are a wonderful beginning to a marriage; the primary reason for marriage, though, is to extend each partner's gene pool. As heir to the throne, Prince Charles knew this perfectly well. Diana, apparently, did not. By the time she had produced the requisite "heir and a spare," the marriage was in deep trouble, and—well, we all know the rest of the story.

It takes a long time to rear children. If there are three or four children, by the time they all are grown and on their own, their parents have been together for thirty or so years. At the end of that time the wife and husband, by now probably in their fifties, take a look at each other without the buffer of children. In earlier ages, with much briefer life expectancy, one or both of them would have died, but today they have a great deal more life to live. It can be a critical juncture in a marriage.

Homeostasis is a powerful force. There are numerous reasons that married people choose to stay together after the children have moved out. Many couples, for instance, have developed shared interests over the years, and when interests are strong enough to foster mutual caring and sharing, the partnership lasts.

Or, if a couple enjoy separate interests, they often work out a way in which, supporting one another, they continue those interests. Again, the marriage remains intact. If separation means a substantial change in social position or financial security, the couple may stay together just because many years together count for something. It's hard for them to change long-established patterns, so they stand by each other as their familiar life rolls along. Every marriage is different, and the husband and wife in every long-term marriage have found ways to maintain the relationship's homeostasis.

Divorce, however, is a pattern-breaker that makes all things new and more difficult.

Any new relationship will demand much more intellectual capital, much more thinking through to understand the other person's needs. It takes years to achieve such understanding in the marriage. Now there are old patterns to un-learn and new ones to learn; at fifty or older, this reconstruction is as daunting as it is stimulating.

Rekindled passion can blind a divorced person to flaws in a new relationship, precipitating a headlong rush into a second, wholly unsuitable marriage. Loneliness can lead a person to settle for an unworthy new partner. Worst of all, perhaps, is a phenomenon commonly encountered by divorced people: the larger cage rejects them because their presence disturbs its homeostasis; they are perceived as a threat to marriages.

Age seems not to be a factor in this circumstance, but gender is. Whereas divorced men are often welcomed into the larger cage—they seem so helpless, somehow, and besides, there's always room for an extra man—divorced women frequently are shunned. It isn't fair, but it's the reality.

Although grown children may be very distressed by their parents' divorce, they usually are occupied with their own families and careers and interests, so that the separation's effect on them is mitigated. When couples with younger children divorce, though, the effects can be devastating.

The family cage has been blown to bits.

The child doesn't understand why, and typically he blames himself. The relationship he had with his parents as a unit has become two individual relationships—that is, if they're lucky. Too often one parent, more often the father, withdraws from the picture and his influence wanes accordingly. Each parent's new cage may lack emotional and financial security, resulting in a lower standard of living. Divorce courts try to do what is in the best interests of the children, but children thrive according to love, not laws.

There may be a new significant other or spouse for one or both parents, with the children enjoined to put aside previous patterns of affection for their father or mother and embrace the new partner. That is, they are expected to respond to a new stimulus in the same way they responded to the old. But the pattern in place is hard to change, and the Brady Bunch is only a silly TV show; resistance to the interloper is almost a given.

With the family cage blasted, it is also common for children to question the values they learned while they were living in it. For example, if a child was reared in a religious home that honored the sanctity of marriage, he may question not only his parents' devotion to their religion but the point of religion itself, and then abandon his own beliefs to move ever farther adrift. So very many things, in fact, can and do go wrong when divorce happens that staying together "for the sake of the children" may seem worth the effort, yet even in the best of these circumstances children sense anger and tension between their parents and react to it unhappily. Half of first American marriages end in divorce. For our children, the only benefit of that statistic is that at least they have a lot of company.

TRIBAL CAGES

HUMANKIND'S BASIS UNIT of existence is the tribe. Survival depends on the group's working together for food, defense, reproduction, and family. Survival supports those members who function most ably in these areas. Tribes cooperate to build shelter, cope with sickness, help the aged, rear children, and develop a tribal religion. Tribespeople tend to reject outsiders as being "other," non-tribal, hence dangerous. Apparently this attitude has been bred into the human animal throughout the millennia.

In Western nations, inclusion is today's ideal. At least we give lip service to it; in practice, however, exclusion is holding its own quite well. On this very day that I am writing, for instance, it finally has become law that openly gay men and women must be allowed to serve in our military. They have been doing so all along, of course, and very creditably, but only if they kept their sexual preference secret. In America, racial and ethnic groups remain for the most part endogamous. (But guess who's coming to dinner.)

Hostility toward "the other," while not praiseworthy, is nevertheless a long-entrenched pattern of human thought and behavior. Left to their own devices, tribesmen typically prefer to breed with and protect their own women. Eventually, though, exclusively endogamous groups confront serious health problems, as continued interbreeding often results in recessive diseases.

Native Americans, for instance, suffer a high incidence of chronic liver disease (different from the liver disease so often associated with alcoholism in these tribes), and among some Pueblo tribes cystic fibrosis and albinism are unusually common. Ashkenazi Jews are subject to a number of recessive conditions rare in other populations. Europe's royals have been intermarrying for centuries, and hemophilia (known as "the royal disease") as well as other recessive conditions still afflict some royal family members.

Even among ancient tribes, the persistence of such health problems would not have gone unnoticed. The tribespeople might not understand the role of genetics, but Nature certainly did. Unimpressed by the flimsy tribal cage, Nature interceded with one of her crude but effective solutions: Might makes right. Here is how it went.

Ocracoke, a strapping young tribesman, is seeking a wife, but he finds the selection of available young women in his tribe unappealing. One is too scrawny to consider, another is cross-eyed (like her mother and grandmother), yet another seems to have some kind of wasting illness that probably means she'll die early. And so forth. None of this augurs well for a successful marriage with many handsome sons. Besides, Ocracoke has had his eye on the lovely Sanibel, a maiden of the hill tribe across the river. He watches from behind a tree sometimes when Sanibel and the other girls come to bathe in the river. Ocracoke must have Sanibel as his bride—but how? How, when each tribe marries only its own? When the tribes are enemies, and contact between them is strictly forbidden, and even small children are beaten for wandering too close to enemy territory? Will our hero heed the laws of the tribe, or will lust conquer all?

It is a dilemma, and Ocracoke must choose. Driven by desire, by his youthful impatience with his stodgy elders' rules, by his dream of all those handsome sons, and by the thought of having to wed his second cousin Alcatraz, who is also his stepsister, he breaks out of his cage and into the enemy's. Swiftly Ocracoke crosses the river, seizes Sanibel from among the other maidens, and bears his prize home.

We will assume that after some intertribal insults and threats and stone-throwing, things went well for the couple. Ocracoke and Sanibel lived happily, surrounded by their many handsome sons and beautiful daughters. Enchanted with their adorable grandchildren, both sets of grandparents set aside hostilities and became friends, and the other elders followed suit. Eventually all the tribespeople came to see the advantages of peaceful cooperation, not least of which was a new generation of healthy children, who as they grew up joined forces with each other, creating one tribe, more numerous and much stronger. After many years the abduction of Sanibel, denounced at the time as an act of war, evolved into the Great Creation Legend of the Ocracokes.

Even today symbolic abductions are commonplace intertribal marriage rituals. On the appointed day, the bride waits happily for the groom or his brothers and cousins to "abduct" her, and after she has been taken, there is no attempt to recapture the girl and she doesn't try to escape. Instead, the groom's parents may visit her parents' home to pay for her. The payment has been agreed to in advance, although there may be a little ritual bargaining. (An friend who worked for years among the Masai says that the greatest compliment she ever received was from a tribal elder who estimated her worth at ten bulls and a hundred cows—an unheard-of bridal price.)

The continued existence of symbolic abductions and similar ritual remnants suggests that despite their lack of knowledge about genetics, tribal people do understand, perhaps through racial memory or maybe just through observation, that mixing gene pools results in hybrid vigor.

Our own sophisticated, complex society is still tribal. We inhabit multiple tribal cages and identify with all of them. You may be, for instance, a Catholic, a Republican, and a New Yorker. You are a member of the international tribe of Catholics but your chief religious identity is with your local parish; you belong to the national

Party, but your political identity is mainly at the district level; you are an urbanite, but your tribal home is your immediate neighborhood. Human beings tend to deal with the world's problems in an abstract way, but effective action happens in an identified group. Rather than impose ideas on the group, people who want to help or influence others usually work within the group, taking into account its patterns and customs. The Peace Corps is a shining example of this method in practice. And many religious missionaries have come to recognize that a judicious mingling of tribal rites and beliefs with their own gains far more converts than outright bans on "heathen" activities. Again, hybrid vigor.

In Conrad's *Heart of Darkness*, Marlow, the British narrator, studies a map of Africa in preparation for his trade mission to the Belgian Congo. Tellingly, he sees the map not as a display of separate African societies but as a color palette denoting which European countries "own" huge areas of the continent. That is, the red areas of the map signify British ownership, the blue French, the green Italian, the orange Portugal, the purple German, the yellow Belgian. To deal successfully with other people, we need to take seriously the collections of patterns that constitute the familial, tribal, and societal cages. Like family cages, tribal cages may remain intact even though tremendous outside pressure tries to divide them. For example, in the nineteenth and early twentieth centuries Africa was colonized; large countries were hammered together by European imperialist powers, European customs and mores were imported into these countries, and generations of African children grew up under European influence. Eurocentrism held sway.

As Africans achieved independence from colonial rule, however, they re-adopted many of their former patterns. They created smaller cages--smaller countries that reflected former times. It was a return to the cages that prevailed before European control. Today Africa includes fifty-four independent nations (its amoeba-like splitting during the past century was quite a challenge to cartographers trying to keep pace), and although the rate of nation-forming has slowed, the trend probably will continue. European control seems actually to have strengthened tribal cages; when the colonizers withdrew, the descendants of those who originally were forced into new ways returned to some of the old ways, reinstituting tribal patterns almost as if the intervening centuries had not existed.

BECAUSE WE LIVE in cages, most of us, most of the time, perceive ourselves as existing in a self-contained sphere moving through time. We don't understand that the world is linear and that each of us is only a sagittal cross-section of the line. The linear world extends endlessly behind and ahead of us, but our perceptions of what the world is all about are local and short-term. These limited perceptions carry over into our lives, relationships, and businesses. Within our business cage, for example, we may be blind to the past and its relationship to the present and future, and therefore become more concerned with tactics than with strategy. Thus many a business has foundered. This short-term perspective, coupled with the natural movement toward entropy, causes us to prefer what is expedient rather than what is good for the society. We drift along in our spheres, largely unaware of the river that is keeping us afloat

As an instance of expediency, I offer "Gilbert's ratio," which involves the number of dollars spent for education in relation to the number spent for the criminal-justice system: judges, prosecutors, defense attorneys, courts—all the apparatus that represents the cost of crime. (Note that I'm not taking into account the psychological or emotional costs of crime and punishment.) My formula is simple: *The higher the ratio of education funding to criminal justice funding, the higher the quality of the cage.*

Americans love crime and punishment; for proof, watch a week's worth of TV. The United States spends more money on criminal justice than it does on educating children. It is more expedient to lock up criminals and throw away the key than to provide the kind of education that will make criminality less attractive in the first place, but incarceration is a short-term and very costly solution.

A well-educated populace, however, is less likely to engage in criminal behavior. By spending the money and investing in the resources that educate our young people, we will save many of the dollars now draining into criminal justice. Educated citizens commit fewer crimes; less criminality means a healthier society. But this long-term systemic solution to a systemic problem will require more than one lifetime, and Americans are notoriously impatient. We want results *now*. Strategic change just isn't very attractive to people who seek immediate gratification.

Caged creatures we inevitably are, yet life in the cage can be in some measure what we choose to make of it. Shall we be solipsists for whom the only reality that matters is our own? I know people who announce proudly that they never watch the news on TV or read newspapers;

purposely, they reject connection with the rest of the world. The earthquake and tsunami that disabled a huge part of Japan's nuclear power system and killed thousands? "Nothing to do with me," says such a person, "and I can't do anything about it anyhow." As our world becomes increasingly complex, perhaps this attitude is understandable, but it angers and frustrates me to hear anyone so blithely embrace his ignorance and so tragically deny his membership in the human community.

How much better, I believe, to acknowledge that no man is an island entire of itself, to gather the strength of mind and spirit that allows us to see beyond our cages, to perceive the possibility of human progress, to dare the escape, and maybe—who can say?—like brave young Ocracoke, to found a nation.

THREE CAGES IN
A ROW

Pied Beauty
Glory be to God for dappled things—
For skies of couple-colour as a brinded cow;
For rose-moles all I stipple upon trout that swim;
Fresh-firecoal chestnut-falls finches' wings;
Landscape plotted and pieced—fold, fallow, and plough;
And all trades, their gear and tackle and trim.
All things counter, original, spare, strange;
Whatever is fickle, freckled (who knows how?)
With swift, slow; sweet, sour; adazzle, dim;
He fathers-forth whose beauty is past change:
Praise him.
- Gerard Manley Hopkins

HOPKINS'S QUEST IS for the numinous, mine is more temporal, but it seems we both eschew the big bright shadowless highways, preferring instead to move toward our destinations along similarly dappled paths. There is where you find the surprises—the "counter, original, spare, strange"—those unique combinations of patterns that make life so interesting. Following these paths, I've visited many cages, and savored the wealth of patterns they enclose. At auctions, for example, and flea markets and antique shops and various gatherings of people, I like to try to winkle out from the jumble the few threads that may be key to a life's tapestry. Here are three snapshots from my wanderings.

SATURDAY MORNING
AT THE GUN SHOW

HUNDREDS OF MIDWESTERNERS are spending this stunning fall day at a gun show in Johnstown, Ohio. The turning leaves are a profusion of color (talk *about* dappled things). The fairgrounds parking lot is crammed with late-model pickups, most of them glossy black, a few in daring shades of red and blue, all of them expensive and well maintained.

The gun show is not a cage for poor people, although most of them seem dressed purposely to mask their prosperity.

There may be a few doctors and lawyers here—the American love for firearms crosses many boundaries—but most are farmers and tradespeople, and many of the men are wearing the same costume: jeans, work shoes, white socks, message-bearing T-shirts, and baseball caps. Broad leather belts with big buckles hold the older men's paunches in check.

The young men, though, are largely broad-shouldered, narrow-hipped, and muscular; they stride or saunter with their shoulders thrust forward, their elbows out, and their hands either in their pockets or out, palms backward-facing. It's an aggressive posture, suitable to a certain kind of young man in a certain kind of cage, but it is as yet only a posture, for there is no question that at this gun show the adult males are dominant and everybody knows it.

As a group, the young women are in Rubenesque bloom, smiling and pretty. Shining hair, minimal cosmetics—these girls probably know quite well how attractive they are; there is no need to improve on nature.

Their mothers are a different story: almost without exception they are tired-looking and too heavy. Some have made obvious efforts to restore their faded looks, and the results are rather sad. Here, for instance, is a woman whose impasto makeup is witness to her losing struggle with age. There is a woman who is wearing clothing two sizes too large to camouflage her obesity. The women trail along close to their men, keeping an eye on their children and saying little. Yet they seem content to be here at the gun show, out of the house on a fine day.

The guns for sale are expensive, costing in the thousands of dollars, and a lot of money is changing hands at the booths. Everybody wants to try the shotguns, all of which are "broken," or open, so you can see they are unloaded. Other weapons, knives and bows, are for sale too, and I'm told that bow hunting is increasingly popular because of movies like *The Hunger Games*. These are modern crossbows, costly, meticulously fabricated, and lethal-looking.

The men handle them enthusiastically, but there are few buyers. Much more is going on than gun sales, though. Rounding a corner, you see a flea market flogging a hodge-podge of wares, from a three-hundred-piece collection of German beer steins to a selection of paintball guns to stacks of Harlequin romances to police badges. There's food: steak sandwiches, hot chicken sandwiches, ham and bean soup, cornbread, batter-fried vegetables. And there's almost-food: fried cheese and corn dogs and funnel cakes.

A shot rings out announcing the start of the fast-draw contest. In this contest the shooter must draw and fire a single-action revolver at a target. Thumbing and fanning are the chief shooting styles; with thumbers using one hand, fanners using both. They shoot wax bullets rather than live rounds, but the wax bullets can hurt. If the shooter's timing is off, he's apt to shoot himself in the leg. "It stings real bad," a man tells me, "and it draws blood sometimes." The holster is rigid, steel-reinforced, so that it doesn't move when the gun is drawn. Participants in the contest represent a sub-cage within the larger cage of gun owners and sportsmen.

At the same time, dog races are being held. Don't expect greyhounds here, or mechanical rabbits. These dogs—purebred hounds and many mixtures—swim across a pond in pursuit of a caged raccoon that is dragged in front of them, then lifted out of their reach, high up a pole. Two dogs can win each race: the dog that swims the pond fastest and the one that reaches the pole and trees the coon first. Or, one dog can win both contests.

The dog races continue all day, accompanied by considerable wagering. The dogs and their owners are having a wonderful time, but as for the raccoon, well,

291

though he isn't physically injured, psychologically he may never be the same.

Judging by looks alone, this crowd is mostly white. Any four male faces, for example, might be variations on the faces of the Rolling Stones. One strawberry-blonde girl, given a Renaissance costume, could pass for the youthful Queen Elizabeth I. As the day wears on, you notice many multigenerational families. Young couples hold hands but that's as far as it goes, at least in public. Small children are generally obedient. People are courteous to one another, friendly and welcoming to me. You don't hear angry shouting. You don't see fighting.

In the midst of all this weaponry, it is a rather surprising civility—surprising, I guess, only to people who are unaccustomed to guns. But it is the guns, certainly, that bring all these people together. Ownership of good weapons for protection and hunting is an ancient essential hearkening back to our earliest days. Even so, on this balmy Saturday in Johnstown it is a little strange to realize that such an atavistic impulse is what attracts these peaceful people.

THE GREEK COMMUNITY in Columbus has built a beautiful Orthodox cathedral, and it is where they present the annual Greek Festival. Over the years, the event has grown steadily. Indeed, you might say that Columbus anticipates the festival with its collective mouth watering. This evening, for instance, hundreds of people are waiting in line to sample dolmades, moussaka, gyros, hummus, spanakopita, and dozens of phyllo-wrapped pastries. (There is much, much more, but I have exhausted my knowledge of Greek food-names.) The mélange of Aegean and Mediterranean flavors and odors brings myth and history immediately to mind: Paris and Helen, Odysseus and Penelope, Zeus and Hera, Homer, Socrates, Aristotle, Hippocrates.

The men are hard at work. They run the festival, and for today they are cooks and bartenders. The women and children serve food, clear the tables, and guide visitors. In rooms inside the cathedral, booths are set up to sell jewelry and clothing and food. Most of these Greek-Americans have the dark hair and eyes we associate with modern Greeks, but there are a few blue-eyed blonds. It's fun to think they are descendants of Helen herself, but their genes probably attest to the long-ago presence in Greece of Viking explorers.

The gorgeous cathedral itself is evidence of the Columbus Greek-American community's affluence. Only people of means can afford this cathedral's intricate architecture and exquisite interior. It is interesting that many of the adults here are first- or second-generation Americans, which speaks volumes for their drive and for their diligence in having acquired the education necessary to join the professions.

I know, for a typical example, of five brothers, sons of a Greek woman who came to this country at fourteen to marry her recently-immigrated Greek husband. The couple was poor, but with the support of their community and their own strong resolve, they thrived. One of the brothers became a lawyer, one a pharmacist, one an oral surgeon, one a research pharmacologist, and one an endocrinologist who directed a huge university hospital. Each has been highly successful in his field. All the grandchildren have entered the professions, mainly law and medicine, and all have married other Greek-Americans.

There is no telling how long the close-knit, endogamous nature of this community will last. No doubt a few of its members have already broken some ties, but in large part the cage walls hold steady. You have only to hear the Greek band playing Greek music while the women wearing traditional Greek costume form circles to dance Greek dances.

As the hour grows later and the wine flows more freely, some of the men dance. They aren't in costume but it doesn't matter, for soon it is clear that they dance not for an audience but for themselves. It is as though these modern American doctors and lawyers and businessmen have transcended the festival and Columbus and perhaps even America, and have merged with antiquity.

These expressions of an ancient ethos confirm and validate a nearly pure cage culture.

The Greek-Americans who work and dance and worship together are observing the customs of their ancestors. Today they appear fully as committed to and observant of their ideals and beliefs as those who came before them. They honor their magnificent legacy, and here in the enormous cage of America, which so often attenuates and sometimes destroys the cultural patterns within it, they hold their culture in trust.

THE SUNDAY FISH FRY

I HAVE DROPPED in on a fund-raising event in Harrisburg, a very small town outside of Columbus, where a people's rights organization is raising money to defend the Second Amendment—the right to keep and bear arms. Two hundred or so people are in attendance, and the fish themselves are frying in a cement-block building where a dozen or more men, women, and children are eating fish sandwiches and drinking Coke or beer.

I don't feel very welcome.

The men wear caps, jeans, and polyester polo shirts that emphasize their bellies. These aren't proud, firm, plutocratic paunches; they are rather soft and unattractive, possibly the result of too much unhealthy food and too little exercise. Many wear pistols on their belts. The men are gruff to the outsider and laconic among themselves. They don't look happy. Their wives are overweight too, especially from the waist down.

For some reason the women are wearing Bermuda-length shorts that only emphasize their large figures. It's kind of nice to see that the older women are frankly older: their hair is gray and unstyled, and it appears they use few, if any, cosmetics. They seem happier than their husbands. Some even smile at me.

In fact almost all the men and women present themselves simply as they are. They tend to stoop. Maybe this is attitudinal, a habit held over from their blue-collar or laborer jobs, where deference to authority is the rule.

Maybe it is part of their culture not to care much about posture. Maybe it is simply a matter of gravity, with their excess weight dragging them down. Or maybe they don't get enough calcium in their diet, which I would guess runs heavily to fatty meats, mashed potatoes, and fried foods, all complemented with soft drinks and beer and salty, greasy snacks. Bone scans might yield interesting results.

I'm isolated and feel purposely ignored, which I suppose is to be expected in this archetypal human tribal cage. My clothes are weekend-casual, but they're the wrong kind of casual. Probably everybody thinks I talk funny. In any case, they clearly have no interest in me except insofar as I may be a government spy, and they resent my intruding on them. I am an outsider. I am suspect. And I am feeling less and less comfortable. My fish sandwich, though way too heavy on the mayonnaise, is the highlight of my day.

These people are undereducated, but I'd bet they proudly trade book learning with their reliance on what they consider their good common sense.

Their cage of like-minded members fulfills their social needs and confirms their thinking. Based on the few words they do say, they seem remarkably unreflective. Suppose a Constitutional exegete should happen by and attempt to explain to them that the ambiguities of our Constitution and indeed of the Second Amendment itself are among the document's great strengths. There is little doubt they would reject him as one of those (in George Wallace's immortal phrase) "pointy-headed innaleckshuls."

A LITTLE MONDAY MORNING QUARTERBACKING

IT WAS A very full weekend. Looking back, I've probably leapt to a number of imprecise conclusions based on too little information. For instance, what do I know about raccoons? Maybe that raccoon at the gun show actually was enjoying outwitting all those hounds at no risk to himself. And there must be more than a few Greek-American slackers, the despair of their families. As for the fish fry folks, for all I know there were a couple of geniuses among them, as well, perhaps, as an FBI mole. Their indifference to me rankled, and caused me to judge them in harsher terms than I should.

The problem, of course, is one I share with almost everybody else: I forgot that I had carted my own cage with its own experiences and habits of thought into these other people's cages. Inevitably, personal perspective colors perceptions. I regret my failings. In my defense, however, I did say at the outset that my descriptions of these events were snapshots—not exhaustive studies, but merely quick takes on the lives of three groups of people at a brief point in time.

For the sake of argument, though, let's say that my observations were mostly on the mark. What I saw was that family is the element that those three cages have in common, but that probably they would find it hard to communicate about other things. Economics, culture, education, and a host of other issues separate them. People who live in such divergent cages seldom perceive each other as mindfully as they ought.

Moreover, nobody, not even a "soul mate," really senses us as we sense ourselves. If, for instance, you ever have been interviewed for a newspaper article, you know that no matter how fair and unbiased the reporter, the story that appears in the paper seems to be about somebody you don't fully recognize. Even your own accurately quoted words ring false to you. That is because the reporter, operating within his writer's cage, has presented the person he saw and heard, who just isn't the you that you know. I often wonder how public figures bridge the gulf between their private, personal identity and the persona that the rest of the world "knows."

Inadequate communication among cages is a serious obstacle when you are trying to forge agreements and achieve consensus for making social change. Suppose, for example, that the only people to survive an atomic cataclysm are the Greeks, the gun-show crowd, and the Second Amendment people. Now it is their responsibility to rebuild a world with a society of laws, institutions, economic and political systems. Maybe they'll reach accord about some aspect of family life, but otherwise they are unlikely to find many things in common; the probable result may be a society at least as factional and fractious as our own.

Actually we needn't contemplate a cataclysmic future in order to envision such an outcome—not when there is a prime example right in front of our noses: our government. Although Republicans and Democrats occupy distinct cages with distinct philosophies of governance, historically they sometimes have managed to compromise in the interests of doing at least some of the work they are elected to do.

CAGE TURBULENCE

IN THE *NEW YORKER*, a Zachary Kanin drawing illustrates an exterminator explaining to a homeowner his system for destroying infestations of ants. The caption: *Instead of poison, I introduce liberal, intellectual ants into the population, eroding the ants' patriotism and causing them to question the authoritarian rule of the queen. Slowly, over generations, it weakens the ants' genetic resolve to the point where they stay in the nest at all times, watching television and writing letters to the editor.*

People who live in dissimilar cages generally live and let live, as long as their cages separate them. When cages collide, however, the clash creates turbulence at the borders of the enclosures. To the great distress of the inhabitants, the turbulence floods like a tsunami into both cages, stirring up intra-cage energy that creates change. Or, as in the case of the ants, change may happen more insidiously, but with equally dramatic results. In America during the past forty years, for example, attitudes between gay and straight cages have undergone both abrupt and measured change.

In 1969 police raided the Stonewall Inn, a gay establishment in Greenwich Village. Their reasoning was that the Inn's patrons were either gay or some other sexual minority, hence, presumably, guilty of moral turpitude and corruptive to "normal" society. Such raids were common practice at the time. They served mainly as an easy form of harassment, as gay people, with reputations and jobs and families to lose if they were outed, didn't fight back. But this time, having, as several remarked, "had enough of this shit," they did. They rioted, and their actions provoked a series of similar eruptions throughout the city. The Stonewall riots of 1969 were the defining event that began the gay rights movement in America.

The movement achieved a second major milestone in 1972, when the American Psychiatric Association agreed to delete homosexuality as a pathology from its *Diagnostic and Statistical Manual of Mental Disorders*. This decision was of immense significance to gay people, and it had far-reaching implications. Few other Americans were familiar with the *DSM* or its implications, however, and only in the past couple of years have polls revealed that with regard to homosexuality, the majority of the majority, i.e. heterosexuals, are willing to live and let live.

These two occurrences—the Stonewall riots and the APA decision—are examples of events that produce immediate and long-term change. Yet despite their importance, for several years the broader American population remained relatively oblivious of the social consequences. In the 1980s, however, with the advent of

AIDS, *everybody* got involved, at a different pace. At first, when no one knew what was causing it, AIDS was perceived only as a "gay men's disease," and straight people could breathe a sigh of relief. Among many, the attitude was "well, it serves them right"; while others were more compassionate, yet not without a touch of schadenfreude. But almost before anybody could take a second breath, scientists isolated the HIV virus responsible for AIDS. Then we realized that in one way or another everyone was at risk for contracting the illness.

The epidemic catalyzed intercage energies, which led to fear-stoked intercage hostility and violence. The signs were everywhere. Fundamentalist preachers convinced their congregations that AIDS was God's punishment for homosexuality. In boxing rings, coaches and cut-men and referees began wearing latex gloves to protect themselves from contact with the fighters' possibly HIV-infected blood. Magic Johnson contracted HIV and immediately retired from basketball. Dentists donned safety glasses and masks and gloves. Some surgical teams, nervous about exposure to blood and other body excreta, arrayed themselves like astronauts prepared for lift-off.

Ordinarily, after high turbulence, there is resolution. In one scenario, each group builds an even stronger cage wall, and at that point energies slow down and life goes on much as before. More usually, however, change occurs in some inhabitants of each cage, and that forges a new relationship between the cages. In the case of AIDS, the hostile factions came to consider themselves not as separate species but as people confronting a common problem. Another enormous change that has happened since the AIDS crisis is the emergence of a new generation, most of whom view all this gay-straight hubbub as a waste of time.

We can see the result now in the homosexual-heterosexual cages. Numerous barriers to gays once forced have come down. Although among many straight people there may still be active resistance to gay marriage, it appears that the term "marriage" is the real stumbling block, rather than the idea of committed civil unions (gay). And millions of people love shows that feature gay couples such as *Modern Family*. It leads me to think that most Americans are more than okay with gay—as long as it isn't "too" gay.

WAR

WAR IS NEVER good, but sometimes it has to be waged. The American colonies had to show King George

III a thing or two. Hitler had to be stopped. Bin Laden richly deserved that bullet through his head.

Apart from the clearly necessary wars, why have we not exercised more care in choosing our battles? Why must we squander lives and treasure over foolish things? Piously we intone that we love peace, yet human history is replete with examples of imprudently chosen conflicts. From the moment that Eve shared the forbidden fruit with Adam, humankind has managed to muck up paradise. Perhaps, though, the urge to wage war is simply an innate component of the human condition. Paradise, after all, would probably be boring.

With seven billion people now inhabiting the earth and using its resources, however, and with that number scheduled to increase exponentially during this century, it is time for us to take a hard objective look at ourselves and, in the interests of survival of our species, try making some improvements.

Turbulence among cages can be quite difficult for the people living inside them. As energy flows back and forth, people may feel conflicted, at war not only with others but with themselves. Because we all live in multiple-cage environments, we may find ourselves embroiled in several battles at the same time.

Warring cages involve different points of view, and different values. It is hard for those caught in the crossfire to determine which way to go. Yes, we must save the whales, but then who is going to save the out-of-work Japanese whalers? Yes, we must seek renewable energy sources, but meanwhile we have places to go, with only petroleum-powered planes, trains, and automobiles to get us there. Yes, child-molesters, having served their sentences, have earned the right to re-join free society—but positively not in our neighborhood. The more cages, the more confusing issues arise. Cognitive dissonance is disorienting and extremely uncomfortable and very important for effecting change.

However awkward things may be for us, global issues now demand thoughtful intercage consideration. We all share the same environment, so that burning high-sulfur coal in West Virginia fouls lakes in Canada, and clear-cutting the Amazon rain forest debases air quality in Colorado. When maquiladoras just across the Mexican border belch heavy clouds of particulate-laden smoke, both Mexicans and Americans develop pulmonary problems. If China's lack of environmental regulations makes its air unbreathable, it isn't only Chinese people who suffer. And of course there is a host of other issues that tear people apart: race, religion, politics—almost any human concern you name can, when cages collide, devolve into casus belli.

They say that repeating the same action over and over again and expecting a different result is a definition of insanity. Possibly. At any rate it's remarkably stupid. And yet, when it comes to solving intercage problems, that has been human history. If we continue mentally and emotionally to be intracage people, thinking only in terms of how events affect us and our fellow inmates, then nothing will change. The sooner we understand that our planet is a global village, the more clearly we will see the pressing need for intercage understanding, and the more likely our chances for intercage cooperation, peace, and survival.

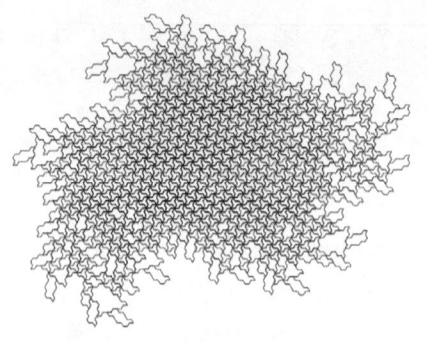

HUMANS, GODS, AND THE CAGE OF RELIGION

"[Religion] is the opium of the people."
- Karl Marx

"Religion is a set of things which the average man thinks he believes and wishes he were certain of."
- Mark Twain

"Religion is all bunk."
- Thomas Edison

"Religion is to do right. It is to love, it is to serve, it is to think, it is to be humble."
- Ralph Waldo Emerson

"My religion is simple. My religion is kindness."
- The Dalai Lama

THERE IS NO dearth of definitions of religion. You might think most people would hesitate to try to define such a complex topic, but choose at random any ten people and you are likely to hear ten different definitions, each delivered with considerable self-confidence. One of the most formidable cages we inhabit is the religion cage. It intersects many other strongholds, including family, tribe, work, community, even national and international cages. I have my own definition, of course, offered with what I hope you'll perceive as suitable diffidence.

A religion is a system of beliefs and practices supporting our relationship with a deity. It is also a set of rules and practices that enable its followers to understand one another and work together. It provides agreed-upon ways to deal with marriage, children, family, sexuality, discipline, and group relationships. It tells us what is expected of us and what we should expect from others. Religion binds people in a structured way of life that helps them live comfortably with each other. Every society and culture has a religion that supports an omnipotent, supernatural deity or deities, gods who typically mete out terrible punishment when mortals violate their laws and customs, and reward the obedient with the promise of an afterlife. Most humans persist in their belief that there must be *something* after death, perhaps because it is so hard, so depressingly final, to imagine themselves finished, kaput, permanently canceled out.

What is it that causes a religion to emerge each time groups of people join to form a society? Why is that religion passed on through generations? Once a religion is established within a cage, it exerts a magnetic force; as it attracts new members, its magnetism becomes more powerful. Religion, with its concomitant sense of belonging, is very hard to resist. A minority may decline to join, and often this group will encounter repression and discrimination.

Early in their lives, most children are taught to pray to a deity. They are patterned to believe that there is a supreme being, omnipotent and omniscient, who protects them from evil, rewards them for being good, and punishes them for being bad. To the society in which the child lives, "good" means doing those things it views as appropriate ways to behave in a variety of contexts; "bad" means behaving inappropriately. There are gray areas, however, where goodness and badness reside in the eye of the beholder. To Dennis the Menace's cantankerous neighbor Mr. Wilson, for example, Dennis is a pest, forever disturbing his attempts to lead a quiet, restful life. To the grandmotherly Mrs. Wilson, however, Dennis is perfectly adorable and no matter what mischief he gets into, she readily forgives him with freshly-baked cookies.

Religion doesn't enter the picture in this comic strip except insofar as there are parameters of behavior beyond which young Dennis never ventures, but that is an important exception. We, the readers of the strip, all know those parameters, and even at the age of five and a half, so does Dennis.

Imagine the outcry if, some morning when millions of us are enjoying the comics with our cornflakes, we should discover Dennis sassing Mrs. Wilson or kicking his dog. We learn limits from our parents, who learned them from their parents, and so on far back into our ancestry. Most people in our society, including those who reject religion, espouse values that are religion-based.

I hesitate to say that this is the result of our still-prevailing Judeo-Christian ethic, both because that may appear to exclude other religions. But believers and non-believers can agree, I think, that religion traditionally has set the ethical rules we generally abide by.

The combination of religion and parental involvement transmits absolute certainty to children and leads to strong bonding within the society. Believers in the same deity support one another. Moreover, they often perceive the deity as exclusively their own, and not as the god of other groups. This perception reinforces the tribal cage so that even overwhelming exterior forces can't destroy it.

The Holocaust, for example, for all its horror, failed to extinguish the Jews' belief in their religion. In tyrannized nations where religion has been actively discouraged, sometimes banned altogether, religious fervor re-emerges when the tyrant is overthrown. Under threat of pain or torture, loyalty to the deity doesn't perish; it merely goes underground until the time is ripe for it to reappear.

Religion was banned in the USSR, but when it collapsed seventy years after the Bolshevik Revolution, tribal cages occupied by people who weren't even alive at the time of repression sprang up throughout the former Soviet Union, and the old beliefs and customs returned intact. According to Fred Weir in *The Christian Science Monitor,* nearly 20 years following the collapse of the atheistic Soviet Union, 82 percent of Russians actually classified themselves as believers in religion. The same study finds, unfortunately, that as religious faith among Russians has increased, so has religious intolerance.

China, on the other hand, remains officially an atheist nation. As an alternative to religion, however, a few years ago the Chinese government launched a God-free "spiritual civilization" program, which extols the virtues of family, loyalty, and diligence. Current government policy does proclaim tolerance of religions, but it is tolerance with an edge: essentially, Chinese people are free to practice religion as long as it does not disrupt the country's social, political, or economic equilibrium. The ambiguity of this caveat leads to continual unease among religious groups, which is probably what the government intends, and rare is the citizen who wears his religious heart on his sleeve. Even so, religion is alive in China, if not particularly robust.

How is it that religion survives long periods of repression and denunciation?

If punishment, ostracism, imprisonment, and even wholesale slaughter of believers fail to destroy it, then religion must indeed be a mighty force that answers a pressing human need. Under the most adverse conditions, when danger threatens people of faith, they still pass on to their children their patterns of religion, thereby ensuring continuity of belief. It is like injecting a genetic virus, the effects of which are experienced through succeeding generations.

Religious groups comprise people who perceive stimuli in similar ways. Out of these sensory, organic perceptions, they create a non-organic existence unique to their group, an existence within which relationships form that endure over time. A man may say, "My father was a Jew. My grandfather was a Jew. Two thousand years ago my ancestors were Jews. Therefore I am a Jew."

It is a construct that, with plenty of help from his parents and his culture, he has created and stored in his brain. It causes him to believe and live in a certain way, and it is a basis for a whole philosophy and practice of living. Others in the Jewish cage share this construct—this genetic virus—and all assume the responsibility for transmitting it to the young. If the transmission doesn't happen, the virus is destroyed. The entire pattern of religious belief can disappear forever.

Although religion's mores and practices are strong, it is nonetheless fragile. It will survive only if each generation carries the gene marker, develops the condition, and passes it on to the children. For example, during the mid-nineteenth century when communal societies were popular in America, the Shaker religious sect attracted a good number of followers. But because celibacy is a chief tenet of Shakerism, year after year the sect's membership diminished. Today, there are only a handful of practicing Shaker communities left.

People frequently fail to recognize their individual significance in keeping their religion cage alive and well. Jewish people, for instance, may go to shul, participate in the services, and join in the Jewish cultural and religious rites, yet consider themselves merely passive players. They understand genetics as something we inherit, but they don't see the pattern virus as analogous, as something that penetrates their psyche and becomes integral to their identity.

The great difference is that the pattern virus is not merely something they passively inherit, but something they must actively perpetuate. It is important to understand this concept, because just as it applies to Judaism, it also applies to other groups: Catholics, Democrats, Republicans, Al Qaeda—any group of people willing to work together to achieve similar ideals and purposes. "Work" is the key word. As Garrison Keillor once pointed out, "Going to church no more makes you a Christian than standing in a garage makes you a car."

IN FREE SOCIETIES, religious groups sometimes find great difficulty in maintaining their members' commitment to their religions' discipline. A conservative faction of the Episcopal church, for instance, has denounced the appointment of an openly gay bishop, and threatens schism.

The Catholic Church often encounters frustration when it comes to dealing with its dwindling, recalcitrant flocks in Western Europe and the United States, where the majority of Catholics tend to pick and choose which of the Church's doctrines to believe or discount, and which strictures to obey or ignore. Pope Benedict XVI, obviously not one to go with the flow, responded by hardening Church policy and focusing on the Church's very successful efforts to win converts in Africa and South America.

Despite the blandishments of TV and movies and the Web, despite the delights of sex, drugs and rock 'n' roll, religion flourishes. The enticements of the open road are so much more exciting than treading the straight and narrow path. It is rather surprising, therefore, that given freedom to choose between the rigors of their religion and the infinity of tempting secular pursuits, so many people continue to cleave to their faith.

Parents pattern their children according to religious precepts, sometimes creating religious cages so strong that the children never even consider escape. Rather, they grow up to inculcate the same patterns in their own children. Sometimes the cage becomes an insular world. In the midst of a society where confusion and chaos reign, the group maintains its equilibrium. When memberships of this kind become well-established and respected, they attract outsiders who may be feeling insecure, who may be seeking some certainty, some sense of belonging, some clear purpose. Such people join the group, "get the virus," and convert to the group's patterns of belief and behavior.

We have seen innumerable groups of "seekers" come and go. Many have been cults of personality, and many of these have wreaked grave harm on their followers and on the rest of the world; Hitler comes immediately to mind. Other such cults, although the damage they inflict is more limited in scope, but hardly less horrific, operate on the same principles: find and attract the vulnerable, persuade them to believe in you and your cause, and get them do your bidding.

As an example of the power of personality, we all remember Charles Manson and his "family." And Marshal Applewhite, the crazed leader of the Heaven's Gate cult, who convinced thirty-eight cult members to join him in mass suicide, after which, they believed, their souls would arrive at an alien spacecraft that was trailing the Hale-Bopp comet, thence to be transported to "a level of existence above human."

Those are cautionary examples. Remember that Hitler called himself Christian, Manson's girls saw their leader as a Christ-figure, and Applewhite believed that he *was* Christ. In a free society, any group can call itself a religion. (As the child sexual abuse scandal involving Penn State University was breaking, I heard a sports journalist refer sardonically, but in many aspects accurately, to the university's football program as "the Church of Football.")

But genuine religious groups, whose intent is to give structure and significance to people's lives, don't prey on the vulnerable. Instead, they are transformative, strengthening their cohorts in belief and purpose and action. These are the groups that not only endure through periods of social laxity, but increase their membership and influence as more and more people find comfort and security among people who reject the outside world's confusion of patterns.

In the United States, Orthodox Jews exemplify this phenomenon. Their numbers have increased even as exogamy rates among other Jews have also increased up to fifty percent. Many American Jews can't accept the constraints of traditional Judaism—the kosher homes, the daily synagogue services, the prescribed and proscribed gender roles. But others find freedom within the boundaries. They welcome the myriad Halachic laws that govern every aspect of ultra-Orthodox Jewish life from Sabbath observances to intimate marital relations. And as they commit fully to these traditional Jewish communities, they establish patterns for their children, infecting them with the virus of traditional Judaism.

While other Jews leave the fold in growing numbers, the Orthodoxy find their own communities swelling with committed participants. Their higher birthrate, coupled with many Jews' desire to return to the essentials of their religion, accounts for the increase. As has been true throughout Jewish history, Judaism's survival will once again rest on the shoulders of the Orthodox.

The major religions are monotheistic but not monolithic. When inhabitants of a religious cage can't resolve their differences, they build new cages that reflect their doctrinal preferences. Inner and external forces, one or both, can shatter religious cages. For example, the Great Schism of 1054 split the Roman Empire's State church into Greek and Latin branches—the Eastern Orthodox Church and the Roman Catholic Church.

In 1517 Martin Luther nailed his *Ninety-Five Theses* to a church door in Wittenberg, and the Protestant Reformation swept Europe during the remainder of the century. In 1534 the Act of Supremacy declared King Henry VIII "the only Supreme Head in Earth of the Church of England," thereby formally confirming Henry's break with Rome and turning England into a Protestant nation.

The Reformation crossed the Atlantic in 1620 in the form of the Puritans, or English Separatists, a deeply unpopular ultra-conservative Protestant sect. This group held that the Reformation simply hadn't gone far enough in Europe. They established a colony notable for its rectitude-unto-nuttiness. Since that time, of course, as the thirteen colonies evolved into the United States, freedom of religion has been one of our identifying doctrines, so sacrosanct that law enforcement and the courts are loathe to interfere even when the practices of some religions are clearly harmful to their followers.

The number of religious denominations in America is hard to calculate, as various breakaway groups, for one reason or another unhappy with their church, form new sects with great frequency. The Baptist and Lutheran churches, for example, include dozens of denominations. Apart from the established churches, moreover, tiny fundamentalist churches seem to spring up overnight. What is just an abandoned storefront today may house a church tomorrow.

Religious organizations pay lobbyists to represent their interests in Congress. For that matter, so do atheist groups. It appears that among humankind the impulse to believe in *something*, even if the something is unbelief, is ineradicable. But why?

WHY RELIGION?

DEAN HAMER, A respected geneticist with the U.S. National Cancer Institute, is the author of *The God Gene: How Faith Is Hardwired into Our Genes.* The science involved is extremely complex, but essentially Hamer proposes that some people inherit a specific set of genes that predisposes them to spiritual or mystic experiences.

The book incurred the wrath of (no, not God) a few of Hamer's fellow-scientists, whose chief objection apparently was that he had chosen to bypass science's accepted route to publication. That is, he didn't submit his material first for assessment by peer-reviewed scientific journals. (The book was a bestseller. Nobody, of course, admitted to harboring any professional jealousy of its success.)

And as you would expect, there was opposition from other quarters. Even though Hamer made it clear he had not tried to prove or disprove the existence of God, to a portion of the public mind his book was sheer devil's work. Some religious leaders and theologians, moreover, protested Hamer's hypothesis that spiritual experience might be the result merely of a physiological quirk, a lucky number in the genetic draw.

My scientific mind finds Hamer's proposal intriguing, but my human heart knows it falls short. For one thing, knowledge is humbling and that *Eureka!* moment is brief. It is axiomatic that the more we learn, the more dispiritingly obvious it becomes how much we don't know and indeed will never learn.

Further, equipped as we are with nimble, curious minds whose reach continuously exceeds our grasp, we know that there is always something more and better, something that transcends our noblest works, that transcends our striving, imperfect selves. The unscalable hurdle is that religion is only in small part a matter of principles. It mostly consists of its congregants, all human, all fallible.

And even though they never fully achieve what they aspire to, they merit great praise, I think, for persisting in the attempt. Perhaps the "why" of religion is that the yearning for transcendence is an element of the human condition. Religion acknowledges our struggles to achieve it, and seeks to teach us how.

This chapter opened with that widely recognized quotation from Karl Marx—words that are abhorrent to believers, mantric to the skeptical. Yet Marx intended them not so much as a judgment against religion itself, but as criticism of a flawed world that cries out for social justice. Perhaps reading them in context will be instructive: "Religion is the sigh of the oppressed creature, the heart of a heartless world, and the soul of soulless conditions. It is the opium of the people."

The cage of religion is massive.

Its membership is enormous. Its scope and influence are universal. Why is it that such vast numbers of people are willingly submissive to an entity empirically unreachable? There are so many possible reasons. Some people find security in a religious group. Others say that religion lends structure and purpose to their lives. Some feel a sense of belonging. Others believe out of custom and habit.

All of these reasons, and many more, may be valid, but none of them is wholly satisfactory.

Granted there was a time, of human history in fact, when it was irrational *not* to believe. But today, when science offers reasonable explanations of phenomena that once could be attributed only to the mysterious workings of deity, our race chooses to go on believing in the divine.

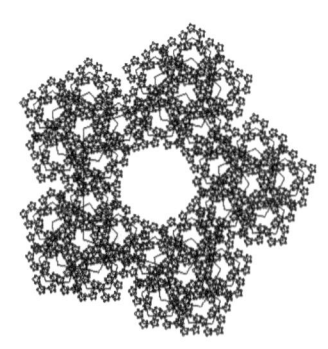

THE CAGES
OF SOCIETY

Just turn me loose,
Let me straddle my old saddle underneath the western skies.
On my cayuse
Let me wander over yonder till I see the mountains rise.
I want to ride to the ridge where the west commences,
Gaze at the moon till I lose my senses,
Can't look at hobbles and I can't stand fences—
Don't fence me in.
- Cole Porter

ON HEARING PORTER'S idyll of the free life, millions of us have wished momentarily that it could come true, then sighed and returned to reality. For we are all fenced in. Even those few real loners among us who seem to have moved beyond the boundaries of tribe, religion, and family still remain enclosed in their self-cages, which can turn out to be just as restrictive, demanding, and sometimes damaging as any others.

I'm thinking for instance of the brilliant aviation pioneer and filmmaker Howard Hughes, who despite his power and immense wealth died friendless, a mental and physical wreck. Ted Kaczynski, the "Unabomber," who devolved from social critic to nihilist, is another gifted loner destroyed from within. But for the rest of us, thankfully, the pressures and requirements of the various cages in which we live or visit are generally acceptable. We acknowledge their value to us personally and to the groups of which we are members, or we just get used to them.

It's important to recognize that the customs of cages other than our own seem appropriate to their inhabitants. As long as the customs do not conflict with our conceptions of basic morality and law, we simply let people in unfamiliar cages live their lives in peace. When they conflict unduly, we take steps.

To the United States of America, for example, slavery became intolerable, so much so that we declared civil war. Warren Jeffs, the Fundamentalist Church of Jesus Christ of Latter-Day Saints church leader, polygamist, and child rapist whom I've mentioned earlier, is now in prison for life.

Probably all of us would like to see something similar happen to Taliban leaders because of their heinous treatment of women. But that is occurring in another cage with customs and laws very different from our own, so we must content ourselves with attempts to exercise moral suasion. Thus far our efforts have been unsuccessful. One reason is that cage inhabitants are so resistant to change imposed from outside.

AIN'T NOBODY HERE BUT US CHICKENS?

A PECKING ORDER. That is, a social hierarchy, is structural to many species including our own. You see it in chickens, wolves, deer, gorillas—any animals that live in groups. Usually a pecking order is established quite naturally, with the strongest and smartest using brawn and brains to take his place on the top rung of the hierarchical ladder. After some skirmishing, the other members of the group settle into their positions.

The pecking order, with a place for everyone and everyone in his place, appears to be a useful structure for living, at least among animals. There will be tensions, especially during mating season as younger males challenge the dominant male for leadership (and droit du seigneur), yet for the most part animals appear to find their location in the pecking order agreeable.

But among humans?

Not so much.

People have trouble accepting that we too are pecking-order animals and that we tend to respond to other people based on where they are in our hierarchy. It seems at odds with democratic values. Nevertheless it is true. As a rule we respect most those who are ahead of us in our pecking order and we accord less respect to those who are below us. The system probably originated in small tribes, where it was valuable because it led to the evolution of strong, intelligent leaders who could safeguard tribal interests, outmaneuver other leaders in war, pursue peace to the tribe's advantage, and not incidentally, produce strong, intelligent children.

That hierarchical system remains in place today. From schoolchildren to university professors to the clergy, the military, corporations, and nations, the pecking order is an established fact of life. Sometimes it is tacitly understood. Among a group of neighborhood children, for example, everyone knows who is the leader, and the followers. And that's not all. Everybody knows which child is smartest or "challenged", which is meanest or kindest, who lies or tells the truth, whose mother is most welcoming, whose house to avoid because the father is usually home and drunk.

All of these aspects of the children's lives, and numerous others—some so subtle as to exist beneath their conscious radar—go into the mix while their pecking order is taking shape. Very few words are spoken aloud about the recipe's ingredients. On the other hand, a grown-up group's pecking order is most often sturdily in place before new members join. A fledgling instructor at a university, his doctorate freshly in hand, understands when he takes the job that he won't be starting as a full professor. There are years ahead, with a daunting number of courses to teach, office hours to keep, committee work to do, administrators to appease, support staff to win over, papers to publish, evaluations to undergo, and professional relationships to forge, before he finally ascends to Olympus.

The new instructor, happy to have the job he worked so hard to get, ordinarily accepts all this as well as his low position in the pecking order, and starts the long trudge upward. Probably he is grateful for the hierarchy. It can show him where to go and how to get there.

The pecking order is a pattern that avoids chaos. At the same time, however, it can be stultifying, and taken to an extreme it can create disaster. Once people are settled within the pattern, great energy is required to blow the pattern apart. This explains why people will go along with things they don't agree with. When a person has accepted the pecking order and the authority of those in charge, pressure mounts on him to do what authority expects him to do.

Probably you are familiar with Yale psychologist Stanley Milgram's experiments during the early 1960s that tested the willingness of people to obey an authority figure who ordered them to do things that clashed with their conscience.

Three people participated in each experiment: an authoritative-looking fellow wearing a white lab coat, a "learner" (an actor), and a "teacher." Only the teacher was, unwittingly, the subject of the experiment. He or she was told it was a study of the nature of learning. This naïve subject, the teacher, was seated in front of what looked to him like an electronics console, with switches labeled from "15 volts" to "450 volts." The learner was seated in another room, out of sight of the teacher, and attached to electrodes that the teacher believed were connected to his console.

The console was a fake; no electricity was used and no one was physically hurt, but the teacher *believed* the set-up was genuine. His task was to read a series of word pairs to the learner and then order him to repeat them correctly. The lab-coated researcher directed the teacher to "punish" the learner with steadily increasing voltage each time he made a mistake. And as the teacher complied, the learner in the next room emitted a crescendo of moans, ending with screams to be let out of the room because he had "heart problems," and then an ominous silence. If the teacher objected to applying higher voltage to a person plainly in pain and possibly dying or dead, or wished to stop altogether, the researcher told him sternly that in the interest of science the experiment must continue.

Milgram repeated the experiment many times. A few of his subjects (very scary people, I'd say) obeyed orders without turning a hair, shocking the learner despite his cries all the way to "450 volts," but most of them protested, and most of them experienced tremendous distress about what they were doing. Some, believing they were inflicting real damage on the learner, did refuse to finish the experiment, but *two thirds or more* of the subjects, their obviously strong moral compunctions notwithstanding, completed it.

Most people cannot bring themselves to violate the pecking order of a cage. The consent of the pecked to continue being pecked maintains the system. Thus we may see entire nations falling into step under an authoritarian leader. And hence in large part the abhorrent deeds of some people during war, such as the massacre at My Lai, the debased behavior of American soldiers at Abu Ghraib, and the tens of thousands of other wartime enormities that have gone unreported.

When Hannah Arendt coined the phrase "banality of evil," she meant that the Holocaust had been the creation not of fanatics and sociopaths, but of very ordinary, that is banal, people carrying out their superiors' directives. Arendt applied the term specifically to Adolph Eichmann, an SS officer and functionary who was a major organizer of the slaughter and whose infamous defense was that he had been "only following orders."

In addition to their willingness to obey authority, people in a pecking order also tend to live up, or down, to expectations. In India, for example, although discrimination based on caste has been officially illegal for more than fifty years, caste plainly remains a social reality.

If, say, a Dalit, or "untouchable," should seek to improve his lot by investing money successfully, buying a plot of land and building a little house, it isn't only higher-caste citizens who may disapprove; people in his own caste are even more likely to condemn him, some going so far as to burn his house and property. Such actions may be partially the result of envy. More accurately, however, they are a matter of identity.

In a hierarchy, your position determines who you are. In other words, you are who the rest of the folks in the pecking order say you are, and your efforts to distinguish yourself cause instability in the order. Dalit or Brahman, it isn't just our own identity that we insist on; it is other people's too, as theirs helps define our cage.

Consider for example a typical state prison for men, where inmates may outnumber guards by a ratio as high as fifty or sixty to one. Overcrowding, gangs, violence, and general squalor, plus many inmates who are psychologically fragile, make these cages very dangerous places to be. Given these conditions, how is it that prison guards, who ordinarily do not carry firearms, usually manage to keep order among their charges?

There are relatively few prison riots and relatively few attacks on guards, and the main reason appears to be that both guards and guarded consent to and sense security in their roles. These separate identities can form very quickly, as Stanford psychologist Philip Zimbardo discovered. In 1971 he designed an experiment to study the psychological effects of being a prisoner or a guard. The Stanford prison experiment is as widely known as Milgram's study of obedience to authority.

Zimbardo built a mock prison in the basement of the University's psychology building, then selected twenty-four Stanford male students as study participants. Twelve would play the role of guards and twelve would be prisoners. Zimbardo was "prison warden." Guards would work eight-hour shifts, but prisoners would remain in prison. The Palo Alto police department cooperated by "arresting" the prisoners, then questioning and strip-searching them before their incarceration, at which time they were presented with a long list of stringent rules of behavior.

He outfitted the guards with khaki shirts and pants from a military surplus store, wooden batons, and mirrored sunglasses that would hide their eyes. The prisoners wore stocking caps, chains around their ankles, and feminine-looking smocks without underwear, and they were addressed by number instead of by name.

The experiment worked only too well. Although the guards were not allowed to harm the prisoners physically for infractions, they devised any number of humiliating torments, such as forcing them to clean toilets with their bare hands.

Within thirty-six hours, one prisoner experienced a psychological breakdown and had to be freed. In the next several days four others were freed for the same reason. As guards became increasingly sadistic, prisoners became increasingly passive, obeying orders no matter how arbitrary or cruel. Zimbardo himself lost sight of his role as researcher and began playing prison warden for real, ignoring the abuses.

Finally a friend of Zimbardo, a woman, came to observe the experiment. Aghast at the conditions of the prison, she told Zimbardo that his experiment was causing serious psychological damage to guards and prisoners alike and that it must end. It was a reality check. Zimbardo terminated what was to have been a two-week experiment on the sixth day.

THOSE PESKY UPSTARTS

THERE ARE ALWAYS members of a pecking order who are so opposed to the dominant members' rule that they drive relentlessly toward establishing their own dominance. It takes more than aggressiveness and energy, however, to move up the hierarchy. The person who would be a leader must be self-confident enough to disregard any cultural imperative to conform to the cage, and resourceful enough to offer improvements to the comfortable status quo.

America is fertile soil for such people. Contemplate the success in so many areas—business, the professions, the arts, the sciences—of countless poor immigrants' children and grandchildren.

The career of President Obama, a man of modest beginnings and mixed race, illustrates the point. Obama survived abandonment by his father and a peripatetic childhood. He endured a somewhat turbulent adolescence during which, as a way, he has said, to avoid issues of self-identity, he used drugs and alcohol.

But Obama's mother had instilled in him the value of education, and that was the path he chose: a degree from Columbia, then Harvard Law, where he was president of the Law Review. Work with law firms followed, and professorships, community organizing, election to the Illinois Senate. Then Obama won his U.S. Senate seat. By the time he had completed just three years in that office, Washington insiders recognized him as a fast-rising star. But, the majority of Americans, when questioned about this freshman Senator Barack Hussein Obama presuming to run for President, could only answer, "Who?"

They would soon find out what the mainstream media wanted them to know.

Some of our presidents have appeared to inherit the office, much like a family fief. Some have been no more than hapless tools of power-brokers. Some have first served as vice-presidents, or chief-executives-in-waiting. A few, like Washington, have had the office thrust upon them. And a few others, often seemingly unlikely candidates such as Lincoln, have been ambitious and astute enough to perceive when the moment was ripe, and quite probably impatient enough to seize it.

Many young legislators have arrived in Washington fired with desire to do useful work swiftly and effectively, only to be balked by the Congress's set-in-stone pecking order. It is well known that Obama chafed at the Senate's arcane rules and hoary habits. (Once, during a particularly tedious Senate hearing, he passed a note to his aide. It read, "Shoot. Me. Now.") Not content to stay in the Senate-cage and work his way slowly up the hierarchy, Obama seized the moment—and broke free.

CAGES ARE UNCOMFORTABLE places for creative thinkers, not just musicians and painters, as creativity by definition continually violates boundaries. Thus, it threatens the cage, which depends for survival on homeostasis. Too much creative thought and activity within a cage can destroy patterns and the cage itself. Perhaps it will be replaced by something better, but probably only by means of a struggle that leaves, literally or figuratively, a lot of bodies in its wake. Comfortably caged people tend to view creative people as subversive of the conventional, the tried and true.

They're right.

We go to great lengths to maintain our established ways of thinking and doing. A creative thinker in our midst may force us to consider our established patterns in a different light, and that can engender discontent within the cage and a breakdown of what we once believed was good order. It is almost inevitable, then, that creative people typically find themselves ignored, ridiculed, or shunned.

It begins early. A friend of mine, a writer, remembers that as a child she would sometimes voice opinions and ideas foreign to the family canon of thought. You might expect the relatives of such a child to appreciate her independent mind, maybe to use the opportunity to discuss things with her.

But no.

"I've never forgotten it," she says. "They were simply dismissive. They'd say, 'Oh, you don't really think that,' when I certainly did really think that, or 'Where on earth did you get such an outlandish idea?' or just laugh and say, 'Elaine, you'd float upstream.' Well, I was a kid, so a lot of my ideas probably were outlandish, but looking back, some of them weren't. I was a senior in college before I could start to believe that people actually might take my thinking seriously."

Creative people don't always play by the rules. They can be hard to get along with. Perhaps because their minds are so frequently roving elsewhere, their teachers may despair, and people who try to reach them emotionally may believe they lack empathy and compassion. Often they are battling demons of their own. Leo Tolstoy's depression made his wife Sonya's life a living hell. Ernest Hemingway warred with his depression and his wives all his life, which he ended with suicide.

The epitaph on Robert Frost's (another depressive) gravestone reads, "I had a lover's quarrel with the world." But Frost's sour relationships with family and neighbors, his notorious irascibility, and the uncompromising nature of his poetry gave the lie to the carefully-tended public mask of kindly New England farmer-poet, suggesting that "love-hate relationship with the world" was more like it, with hate usually in the ascendant.

Those whose creativity falls within the patterns of the cage are free to pursue their passions as long as they don't venture too far beyond accepted boundaries. Thomas Gainsborough, for example, a favorite of King George III, profited handsomely from royal patronage. Although Gainsborough grew bored with painting portraits of the nobility, he didn't grow bored with accepting their money. Pablo Picasso earned international acclaim very early in his career, and he remained extraordinarily productive until his death. Picasso probably was the richest artist in history, and by most accounts a happy one.

That English professor friend of mine once found herself stranded in an east-side Cleveland bar with a Hell's Angel who had been introduced to her as having recently completed his prison sentence for murder. He suggested to her that they probably didn't have much in common. Wishing (at all costs!) not to offend an Angel, she thought fast, and offered that indeed they did: they both were considered outsiders to be feared and loathed, but secretly they were admired and envied.

The Angel weighed this novel idea carefully, smiled (revealing one incisor of gleaming gold, the other embedded with a tastefully small diamond), and allowed as how she might be right. Then they shared a friendly couple of beers.

I repeat my friend's little adventure in the underworld by way of suggesting that America has yet to resolve its ambivalence toward intellectuals and artists. Money gets in the way. Who garners more respect from our population, the research chemist at work on a project that, twenty years from now, may result in a cancer cure, or the barely literate seven-foot, seven-figure-salaried basketball star? Whose voice is more familiar to our children, Kathleen Battle's or Jay-Z's?

Across Ohio, high school football coaches pattern themselves and their teams after The Ohio State University football coach and team, and high school football players are their school and communities' local heroes. Similar circumstances obtain in many states, and the objection is not that young athletes get so much glory, but that young intellectuals get so little. Which path to success appeals more to a high school boy, the years of further schooling, the grind that leads slowly to the Ph.D. and the professorship, or the swift rise to quarterback at State, the Heisman, the first-draft NFL choice?

VISUALIZE A SOCIAL pecking order as a triangle, with a power broker at the apex. How does he get there? Then picture one of those human pyramids that cheerleading squads execute. The apex is the most precarious position, with minimal room for error. It is equally important, therefore, once a person is established at the top, to know how to stay there.

Power brokers ascend through a variety of means: popular acclaim or election, superior intelligence, military conquest, sometimes by simple brute force, sometimes merely by being the right man or woman at the propitious time and place. In any case, the occupier of the pinnacle, particularly if he has risen through force, attempts to create his own society by modulating values and mores inherited from the previous culture into a different system.

Saddam Hussein, for example, trumpeted Iraq's glorious past as the cradle of civilization while at the same time employing decidedly uncivilized methods to reinforce his power as dictator. The Shah of Iran took the twenty-fifth year of his reign as the occasion to produce a month-long extravaganza the likes of which Iran had never before seen: a nightly water-and-light show of spiraling fountains and gigantic glittering chandeliers strung high across all of Teheran.

Accompanied by a media blitz of pro-Shah propaganda, the spectacle was meant to celebrate the Pahlavi "dynasty" as the latest in a twenty-five-*hundred*-year line of Persian royalty. Never mind that the Shah's father, from whom he inherited his crown, had begun as an army stable-boy who then bullied his way into power, or that the Shah himself ascended the Peacock Throne only by the grace of the CIA, or that he occupied it at the bidding of the U.S. government.

And let's not forget that paradigm of self-aggrandizement gone mad, His Excellency, President for Life, Field Marshal Al Hadji, Doctor, Idi Amin Dada, VC, DSO, MC, Lord of All the Beasts of the Earth and Fishes of the Sea, and Conqueror of the British Empire in Africa in General and Uganda in Particular. Oh, and also the uncrowned King of Scotland.

A power triangle can seem static, but in reality it is full of opposing energies. Energy streams downward from the power brokers and pushes upward from cage members trying to advance their positions. Once the power structure has taken firm shape, most of its members recognize no other kind of world. They keep pushing toward the apex, and those who get closest to it are the ones ready to take over as attrition of one kind or another (retirement, illness, disgrace, coup, assassination) claims the top people. New power brokers take the reins and, with varying degrees of tweaking, maintain the system. New arrivals in the society will be patterned from birth to work within the triangle, slipping easily into their pre-set roles.

The U.S. Army Officer Corps illustrates a smoothly-functioning power triangle. At its base are the West Point plebes, just starting out on a college education that includes a clearly defined military career path. Upon graduation, they are commissioned as second lieutenants—the lowest officer rank. Before them lies the possibility of successively higher rank and concomitantly increasing power: first lieutenant, captain, major, lieutenant colonel, colonel, the several generalships, perhaps a Joint Chiefs of Staff seat, possibly even a Presidency—the Commander-in-Chiefdom, the universally acknowledged top of the heap.

POWER IS INHERENTLY corruptive and it is true that absolute power corrupts absolutely.

Power is also addictive. Given a taste of it, most people crave more. But like all addictive substances, power is treacherous.

The gravely addicted, his lust for power having blinded and deafened him to law, morality, common sense and common humanity, will stop at nothing to maintain his habit until finally it leads him to a terrible end—failure, disgrace, even death. Human history is replete with examples, but we need look no further back than our past century to find more: Mubarak, Khadafi, Stalin.

Today we are living in insecure times. War, somewhere, is always being waged or seems imminent. We have narrowly averted a second Great Depression but the danger still lurks. Unstable nations have nuclear capability. The threat of terrorism is a constant. Millions of Americans have lost their jobs and their homes. The enormous gulf between rich and poor steadily widens, swamping our economic middle class. The more insecure a society, the riper it is to choose as leader someone whose chief qualification is what appears to be an unwavering faith in his ability to deal with whatever happens.

Again and again we have seen insecure people in insecure times blindly follow plainly crazed leaders, and in the process entire societies turn mad. Germany, one of the most culturally, intellectually, and scientifically advanced nations of the twentieth century, unquestioningly accepted Hitler's leadership and followed him into war and unspeakable atrocity. Despite (and because of) its resurgence into European dominance, and despite its official renunciation of its shameful past, many thoughtful people still regard Germany's motives with misgiving.

There is considerable irony in the fact that insecure people may elect to follow deeply flawed, even psychotic, leaders. The leaders themselves are insecure. Why else would they succumb to the need for so much power? Paranoia underlies the patterns of those who wield great power and relentlessly seek more. Such people are hypervigilant. Their antennae are out, continually sensing danger to their power base.

And when a threat is perceived, the power player moves immediately to crush it, even though the danger may exist only in his own fevered mind. Richard Nixon's notorious "enemies list," which included hundreds of names changed constantly as circumstances change, illustrates the kind of warped thinking that paranoia generates.

SOURCES OF POWER

A SUCCESSFUL POWER broker will go along with the system already in place and attempt to modify it gradually, or gain the support of the military and then make change, usually oppressive change, rapidly. There is always the possibility, however, that those who wish to maintain the old patterns will rebuff a potential leader, or that conflict will arise between the leader and the members of a highly patterned society.

Not so very long ago, "Potomac fever" meant malaria, a disease carried by the hordes of mosquitoes inhabiting the mists and fogs arising along the Potomac River. Potomac fever infected and in fact killed so many Washingtonians that European governments considered the city a hardship post and offered their diplomats hardship pay to serve there. Today Potomac fever is a metaphor for political power lust, as infectious as malaria and often as deadly.

For example, most Americans support the Presidency as an institution, but they resent an "Imperial

347

Presidency." They believe in the checks and balances that our founding documents guarantee, and fear the erosion of basic constitutional rights and the ascendancy of only one branch of government.

Public opinion notwithstanding, however, our presidents have sought and sometimes simply appropriated powers that the Constitution explicitly or implicitly denies them: the power to declare war, the power of the purse, the power of immunity from legislative oversight. In March 2011, for instance, Barack Obama ordered an attack on Libya without consulting Congress. When they returned from recess, Congress objected, but not very much.

A free press is essential to democracy. As we were taught in sixth or seventh grade, a free press is the "watchdog" that keeps politicians and institutions honest. Therefore a power broker seeking ever greater powers must control information. To prevent people from blocking his rise, a leader must keep them in the dark and unable to communicate with each other.

Before this century, dictators found such repression much easier to achieve. As we have seen in the Middle East, the explosion of new media has created considerable hardship for dictators wishing to maintain their supremacy. Those leaders, who with relative ease, once imposed their autocratic patterns on society have seen those patterns turned inside out and their worlds turned upside down.

In Western society, however, the most insidious approach to control of free media is to persuade people that information transmission itself is the cause of problems. "It's the media's fault!" is the cry of politicians and citizens alike. When the press speaks truth to power, attacks on the media (those "nattering nabobs of negativism," as Vice President Spiro Agnew used to call them) increase. "It's not our fault the country's in such a mess," the power brokers protest; "it's the media's fault!" The true culprits, people start to believe, are those who report the facts. A smart manipulator of societal patterns who can sell the idea that the free flow of information is dangerous positions himself to acquire ever greater power, and once the media are muffled, he can violate patterns with impunity because so few people know what he is doing.

The society that fails to guard its freedom of information risks all its freedoms. When power-hungry people impede the information flow, societal patterns founder. Today in America, too many people are too ready to opine that the media are overzealous. It is true that careless reportage results in misinformation. Reputable media should take pains to avoid such mistakes, and should they occur, apply harsh sanctions on the offending reporter.

Far more dangerous, however, is the important story *not* covered, the lie *not* revealed, the bad decision *not* exposed. At present all our major media outlets are corporate-owned. Perhaps this is one reason that in recent years there has been so little genuine investigative reporting. Maybe I'm wrong, but with respect to freedom of information, it's hard to believe in corporate benevolence.

After the Constitutional Convention that successfully produced our Constitution, and after his election as first President had been secured, George Washington wrote to a friend, "The establishment of our new Government seemed to be the last great experiment for promoting human happiness by reasonable compact in civil Society."

The great experiment is still unfinished.

Testing continues to take place in our enormous living national laboratory. Inevitably mistakes happen. For example, the Volstead Act, or Prohibition—the Eighteenth Amendment of 1919—had the unintended consequence of turning into instant criminals millions of American citizens who refused to allow the government to dictate their drinking habits.

Moreover, the Amendment was a great boon to organized crime. Among the public, bootlegging became an accepted way of doing business. Regardless of the murder and mayhem that attached to trafficking illegal liquor, the big bootleggers were admired and even glamorized. It wasn't until 1933 that Prohibition was repealed officially, but by then what little force it had ever exerted had dwindled to nil, anyhow.

Prohibition failed because Americans rejected the government's attempt to intrude moralistically into their personal lives. A century later, however, and little more than a month after Al Qaeda's attack on the Twin Towers, Congress pushed through the Patriot Act. Its provisions dramatically extended law enforcement agencies' ability to search telephone, e-mail, medical, financial, and other records, and increased their power to detain and deport immigrants suspected of terrorist connections.

It also broadened the definition of terrorism to include domestic terrorism. Except for a few thoughtful civil libertarians, in the confused aftermath of the 9/11 attack, not many people questioned the scope and constitutionality of the Act. Perhaps more might have objected had they realized that "Patriot" actually is part of an acronym: the USAPATRIOT (Act), that is, Uniting and

Strengthening America by Providing Appropriate Tools Required to Intercept and Obstruct Terrorism.

How Orwell would have loved that.

Possibly the least noxious of the Act's results is the extraordinary sight of shoeless airline travelers shuffling through metal detectors and undergoing full-body X-rays and gropings by Transportation Security Administration workers. If any space aliens are watching, they're probably hugely entertained. The Patriot Act begs the question of what is more dangerous, open warfare, or violations of our cherished patterns of civil liberty?

PATTERNS OF LEADERSHIP
AND CHANGE

COMMUNITY LEADERS USUALLY assume their communities' patterns, and their responses mirror the communities' responses. Electing new city council members or county commissioners seldom creates much change in a community; generally speaking, they are merely replacement parts. Inertia is a mighty force, so that even when enthusiastic reformers are elected, they mostly end up following the patterns already in place. It takes a violent upheaval to overcome inertia and modify patterns.

The chief role of the cage is to maintain the status quo. Each cage usually is run by an oligarchy or individual person who calls the shots. Each of the infamous Five Families of New York, for instance, has one acknowledged boss. Most American small town governments are oligarchic; sometimes the leaders trade positions and sometimes new members join the ruling structure, but patterns of thinking and decision-making remain constant.

In a stable system, the governed accept governance—indeed welcome it—because that is what they are patterned to do, and because the status quo is working; that is, it keeps the cage safe. The governing citizens of a little Southwestern town populated mainly by Hispanic Catholics who are Democrats, for instance, are very likely to be Hispanic, Catholic, and Democrats. From single-celled organisms to complex government bureaucracies, all living bodies seek homeostasis. If an amoeba finds itself in a potentially lethal environment, it will ball up, secrete a protective membrane, and become a microbial cyst, remaining dormant until it senses that conditions have become more favorable. Then it takes up its orderly amoebic existence again.

If one of our three branches of government appears to be usurping powers that belong to another branch, howls of protest erupt, interminable arguments ensue, new laws are hastily written, and general chaos reigns until balance is restored. Yet for all its virtues, homeostasis can be self-defeating. If the environment changes, homeostasis can preserve a status quo that is no longer effective and that can undermine a society. If that amoeba stays curled up in a ball too long, for example, it will die.

Openness to new people and new ideas may lead to a sense of risk and uncertainty. Our first impulse is to perceive a loss as negative, but often further thought reveals that it is not, that actually it is a positive change to be embraced. If not, the society may die away because it can't handle the loss of worn out patterns.

In a TV commercial for Progressive Insurance, "Flo," the young woman who apparently manages the Progressive store, suspects that "Flobot," a robot who looks much like her, may soon in fact replace her. So Flo disables Flobot by slyly removing her battery. Would that safeguarding one's job were so simple in this age of rapid technological change.

Once we have succeeded in creating robots in our own image, shall we then be as gods? Don't count on it. We'll just be out of work. To survive, people must be ready to repattern quickly. The schools should prepare our children to do that. As I've said before, we should focus on teaching children to think rather than how to pass tests.

Instead of teaching them the One Right Answer, we need to teach children to assess new situations, ask cogent questions, and respond nimbly. But much of our education system is mired in outmoded methods.

Children need to learn to think critically, utilizing higher order thinking skills, such as synthesis and evaluation. The focus should be on skills, not discipline. In many classrooms the relationship between teacher and students is adversarial. What a waste of time that is—time that would better be spent developing intellectual discipline. Critical thinking skills will serve students long after the tests have been taken and new issues must be resolved.

EVOLUTION OR REVOLUTION

TODAY THERE IS genuine revolution, with its attendant violence and confusion and stormings of Bastilles, among Mideast nations. At this time in our own country, we are witnessing what is thus far a pale copy of real revolution.

In Western nations, however, change is usually evolutionary rather than revolutionary. People in power work behind the scenes to effect change gradually, because citizens share a well-established system of societal patterns that counterbalances mutinous impulses. That overarching social perception is integral to the structure of government, so that government officials come to power equipped with fixed ideas about how to maintain the workings of their culture.

We have rules already in place that determine how political leadership should function and how our society should act. Despite, for instance, Americans' collective suspicion that most politicians lie most of the time, when they are presented incontrovertible evidence of lying (as in Watergate), they react strongly. "Power is the greatest aphrodisiac of all," Henry Kissinger once intoned, and despite the well-known attraction of young women toward alpha males, when Bill Clinton's furtive little fling with Monica Lewinsky became public, shock and anger ensued among the public.

"Oh, my God!" cried horrified citizens. "He did it *in the Oval Office!*" (I enjoyed pointing out that he couldn't very well have been expected to do it in Motel 6, now could he?) It wasn't Nixon and Clinton's actual sins that so offended people; it was the damage done to their image of how American presidents ought to behave.

It would be well to remind ourselves occasionally that many of the rules now in place to protect our society were enacted by people who were waging revolution against an English monarch's usurpation of their rights. The

Declaration enumerated the King's transgressions and the Constitution set them right.

Americans accepted the new rules, incorporated them into their patterns, and taught them to their children. The rules became part of the individual and cage matrices, embedded in the social fabric. Today, unless violent change is taking place, people striving for power will try to acquire it by working within established rules. Change thus evolves, and because most effective power players are essentially cautious, evolution may require generations.

In the process of altering a societal cage, power brokers themselves change—for better or for worse. With some, success has a mellowing effect; for others, success only escalates their desire for more power. Fortunately for us, once the Revolution was won, America's young, rebellious founders continued staunchly to stand by their principles. They themselves evolved into serious statesmen, and their countrymen evolved with them into thoughtful participants in a new republic that, in spite of its flaws and its mistakes, would indeed become a beacon to people all across the earth.

America still is evolving, still a great experiment.

Every generation is in some ways different from the preceding one. New leaders and new issues appear; circumstances change fast and beliefs can change equally fast with them. How silly it seems now, for instance, that gay men and women used to be barred from military service—and yet that was little more than years ago. But no matter how many changes America has weathered, no matter how often its principles seem more honored in the breach than the observance, that original matrix remains at its core.

That is, we the people of the United States, probably unlike any other nation, stubbornly persist in our efforts to form not merely a perfect union, but a *more* perfect one. Although a few grumpy grammarians still deplore the pleonasm, and granted that the goal is unrealistic, there is a youthful idealism to it that continues to inform our country, and that I find wholly admirable.

THE RATE OF CHANGE

THE RATE OF current change is related to the rate of previous changes. Societies that have changed slowly in the past usually continue to change slowly. Societies with patterns of rapid change usually go on changing rapidly until change factors cease. Then, patterns that evolved during the rapid change period will become fixed, as the society slows its rate of change and consolidates the changes that have taken place.

Compare societal change to waves. When there is no wind, the water is calm, but a light wind or a slight difference in temperature creates action in the water. Ripples form, then waves. A storm comes in and kicks up high waves that can do damage, and sometimes a hurricane destroys communities.

Our long-term patterns of perception, belief, and action are like long, quiet ocean swells. These patterns are genetically advantageous for our survival. They must be; for in only two hundred thousand or so years, a tiny fraction of the age of the earth, our species has succeeded in colonizing the entire planet, surviving numberless calamitous events. Not even cataclysms like the simultaneous tsunami, earthquake, and nuclear meltdown that Japan suffered in 2011 have stopped us. Such enormous upheavals are critical at the time and place they happen, but over time their impact diminishes as familiar long-term patterns re-emerge and restore balance.

If a disaster, such as an enormous asteroid strike or a lethal virus to which we have no immunity, were to affect the earth's total environment, however, humans would be forced into pattern change. Either that, or die off like the dinosaurs.

Stable societies are essentially conservative, reluctant to generate change, and a stable populace views change with suspicion and distrust. But an unstable society, its energies roiling, is ripe for change. This period of turbulence can be dangerous. As fright and confusion make people vulnerable to demagoguery, evil leaders may take control, imposing repressive patterns on a society. On the other hand, some populations who survived extreme turbulence have rebounded with renewed, refocused energies.

In the late Middle Ages, for example, western Europe was generally stable, its hierarchical social and political systems well established, its economies prospering. Even peasants could be upwardly mobile, and a strong middle economic class had begun to develop. Then, in the form of bubonic plague, catastrophe struck.

One of the most devastating pandemics in human history, the Black Death, peaking from 1348 to 1350, killed thirty to sixty percent of Europe's people. Terror and helplessness accompanied the plague. Famine, broken families, vanished towns, labor shortages, social unrest, political and economic turmoil, and loss of religious faith followed in its wake. It took a hundred-fifty years for Europe's population to restore itself to pre-plague numbers, yet during that time the Renaissance came into full flower. Science, medicine, literature and the arts, travel, exploration, statecraft—all of these flourished and enriched the world. Rising out of the nightmare horrors of the Black Death, the Renaissance was not only a marvel of human resilience, it was a triumph of the human spirit.

THE ENGINES OF CHANGE

CAGES ARE CLEARLY not immutable since they respond to forces within and without. If there is too much imbalance, violent explosion or implosion can result. When the combined pressures of racism, poverty, unemployment, and crime reach critical mass in America's inner-city cages, their citizens may riot. That's explosion. When greed, mendacity, recklessness, and unbridled arrogance undermine the foundations of America's financial structure, Wall Street may collapse.

That's implosion.

Like the people who inhabit them, societal cages go through developmental stages. In a young society, patterns are new and often in conflict. Positions in the pecking order are not yet in place. Changes happen quickly, and just as quickly give way to other changes. As the society develops into mature middle age, it becomes more set in its ways. It has established strong patterns. But after many years those patterns become encrusted with age. The society fears and rejects change. Society is out of touch with the rest of the world, just hanging on until it finally self-destructs. The passages from birth to death may take centuries, but they are inevitable.

Rebirth, however, is possible.

Though some societies vanish completely, others, having learned from their mistakes, rise to power again. Once it was true, for example, that the sun never set on the vast British Empire. When that empire collapsed, it appeared that England's global authority was over. Yet today, having emerged battered but triumphant from two world wars and having acknowledged the need for cooperation rather than dominance, England is an influential player in the world community.

Meanwhile, Germany has risen from the wreckage and shame of its Hitler years to become the Eurozone's only economically stable nation. Other countries hope fervently that Germany has learned it must restrain its historically aggressive nature.

In an organism or a society, entropy—movement from order to disorder—is inevitable. Past a certain age, one's own body experiences it. One by one, and with increasing speed, the body's systems falter and break down. Medical treatment can slow the process. Maybe, for instance, you're so happy with your nice new zirconium knee that you toss away your cane and go out dancing.

That's great. Have fun. Just be careful not to slip, or the next item on your parts replacement menu will be a ceramic hip.

Half a century ago, a rather conservative friend of mine attended a showing of *Hair*. When the final curtain fell, he rose from his seat and declaimed, "Nobody over twenty-one should be allowed to see this play!" In the Sixties, America appeared to be flying apart. Young people, and many not so young, were challenging everything: politics, music, art, literature, public life and family life. Throughout that decade, college students staged sit-ins on campuses and occupied administrative offices.

Radical groups such as the Black Panthers and the Weathermen formed. Young men, often with their parents' encouragement, fled to Canada to avoid military service. Folksingers decried social inequalities. The Rolling Stones joined the British Invasion, their edgy rhythms and lyrics fueling rebellion. (Mick and Keith both still touring; the Stones are still going strong. Good on you, guys.)

Generations were at war with each other. It seemed that every American institution was under siege. These days graybeards may bemoan the continuing influence of the Sixties. "Oh, the sex. Alas, the drugs. Dear me, the rock 'n' roll". Here's the real scoop: the Sixties were *fun*.

But the era was too hot not to cool down. Its turbulence waned. America survived with most of its ideals and institutions intact. Some changes dating from that period have been incorporated into the fabric of American life: the Civil Rights Act of 1964, for one, while other changes were discarded. The young Turks cut their hair short. (Oddly enough, today you're likely to see long hair adorning poor young white men in rural America. These were the very folks who once denounced it loudest.) Some youth from the sixties opted to join the system and try to change it from the inside. SDS founder Tom Hayden, for instance, served in California's State Assembly and State Senate. Others just threw in the towel and became respectable citizens with regular jobs.

When confusing, sustained change is taking place, people in its midst may fear the end of the world is nigh, or at least the end of their world. Unless a leader and his government have been violently overthrown, it usually isn't true. And even in that extreme event, some kind of stability eventually will be restored and people will carry on, integrating new rules and customs into their lives. For example, when *Hamlet* ends in carnage and the stage is strewn with dead bodies, it seems impossible that Denmark will ever be the same again. And it won't be, not precisely the same. But enter young Fortinbras of Norway, ready to take command and re-establish order.

Ordinarily, though, when outworn ways need replacing, change comes about gradually: birth, maturity, decline, death. It is the nature of entropy. Tennyson's Arthur understood the cycle. Close to death, attempting to comfort his devoted Sir Bedivere, Arthur reminds the grieving knight that "the old order changeth, yielding place to new." Yet rather than succumb to sorrow, we might keep in mind the potential for rebirth. *Rex Quondam, Rexque Futurus*: those were the words carved on Arthur's tomb.

The Once and Future King.

Or so it is said.

THERAPEUTIC APPROACHES TO PATTERN CHANGE

"It is not the strongest of the species that survive, nor the most intelligent, but the one most responsive to change."
- Charles Darwin

WHEN YOU SPRAIN your ankle or break your arm, it's obvious you need medical attention and you don't usually hesitate to seek it. When your psyche is out of kilter, though, you're very likely to balk at asking for help. For one thing, many people continue to stigmatize mental malfunction. Further, they are deeply suspicious, even fearful, of mental health practitioners. The endless joking about psychiatrists attests to this fear, as a typical human defense against something that frightens us is to make fun of it. And too often, people suffering from depression or anxiety or some other condition that destroys their joy in living blame themselves. "I could pull out of this depression if I only had enough willpower. I'm weak," they tell themselves, all the while sinking lower into despair. Or, "Lots of people have bigger problems than I do. I'm taking myself too seriously." Or, "I must be nuts." None of these self-diagnoses is helpful; all of them result in feeling worse.

It is quite natural to fear revealing our foibles and weaknesses and perhaps shameful behavior to a stranger, for that leaves us vulnerable. Although we can see that some of the patterns we have fallen into are self-destructive, they are so deeply grooved that they have become integral to our identity. Disappointed though we may be with ourselves, they are nonetheless *our selves*: private, precious, and mightily resistant to change.

I know from personal experience how hard it is to change patterns. Many years ago I developed a serious free-floating anxiety that I couldn't control. I invested eight years in Freudian analysis, completing four to five sessions per week, fifty weeks per year in an attempt to get to the issues that underlay the anxiety.

For a long time I didn't realize that my distress resulted from my aberrant perceptual patterns. As a means to protect myself from hurt, I had erected too many barriers against genuine intimacy with others and against intimate knowledge of myself.

What a struggle it was.

But with my psychiatrist's help, I finally was able to reach far down to the sources of my destructive patterns, to understand their impact, and to change. By the end of the eight years, my symptoms had been alleviated to the point where, no longer plagued by overwhelming anxiety, I could lead a full and productive life.

Those eight years were invaluable to me, and I am grateful to have spent the time and effort. Even so, not a day passes without something eliciting a response that causes me a twinge—a fear of regression, a sense of discomfort—although my treatment ended more than fifty years ago.

"What is past is prologue," Shakespeare wrote. And people who disdain to examine their own history or fear to plumb its depths are indeed condemned to repeat it. In my case the unexamined life was not worth living, and I knew that things must change but I didn't know how to change them. Fortunately I sought professional intervention to help me recognize and deal with the patterns established in my childhood.

I have a friend who is not so lucky. Cecil's entire world is dark. He is bright and creative, yet he perceives all stimuli as ominous. Point out to him the beauty of a clear blue sky, and he will see only the tiny gray cloud miles away, and the possibility of its developing into a tornado. A minor mishap with his car, a fender-bender that would only annoy most people, is catastrophic for him.

Everything exacerbates his nearly lifelong depression, a condition so oppressive that it has stunted his emotional life and barred the way to consistent success in his field. Cecil's work is acclaimed, and some of it has won awards, but his whole psyche seems focused on what he perceives as the inevitability of disaster and failure. It is heartbreaking to see so much of his creative energy consumed by neurosis. A therapeutic solution to his problem will require him to change his ingrained perceptual patterns and learn how to respond in new, positive ways. That will mean a long, complex process. Yet, given the heavy pall of depression that has shrouded Cecil's life, both undermining his success and protecting him from hurt, I doubt he has the will to engage in it.

Today, because of the time and expense involved, few people turn to classical Freudian analysis. Health insurers won't pay for it. New techniques have evolved, however, and new drugs, that clearly can speed up the process. But no matter the technique used, eventually the patient and the therapist must get down to the patterns in place and the need to change them. It will take a long time to modify those patterns. Here the therapist must tread carefully, lest the patient's fear of change cause him to leave treatment too soon. A friend suffering the effects of childhood sexual abuse by her grandfather told me what she experienced when her therapist began speaking seriously about pattern change.

"For three or four months we'd been discussing the consequences of the abuse," she said, "and I believed I'd arrived at a good understanding of myself and my patterns of behavior. I was feeling a little smug about it, really—like a good student. Dr. Morris had mentioned the need for change several times, but we'd pretty much skirted the issue. Or I had. But there came a day when he got down to essentials: 'Sarah, you need to change. We have to start talking about change.' Well, I thought I was ready for it. My intellect told me that change was required. My body told me something else. As we discussed my need to change, I realized that I had gone rigid. My heart raced, and cold sweat was trickling down my back. I was actually, literally, wringing my hands. *Change.* Me? Enter uncharted territory? Leave the old map behind? Fend for myself in the wilderness that could be my future? At that moment, 'change' was the most terrifying word I had ever heard."

THERAPY AND INTELLECT

WITH GUIDANCE FROM their therapists, many patients find it fairly easy to intellectualize their problems and their patterns, and it is tempting, therefore, for therapists to continue to deliver treatment at that intellectual, cause-effect level. But intellectual understanding is only a start, as patterns are established subcortically and cannot be reached through the intellect alone.

For instance, I know a woman whose very attractive face and figure might be the envy of most other women. Yet when this woman pictures herself, she sees only flaws. Her subconscious image of herself is so distorted that, despite the heads that turn when she enters a room, the admiring glances and the multiple compliments, she can't believe the evidence. She says, "I look okay, I guess [*Okay?* The woman's gorgeous!], but you know, whenever I catch an unexpected glimpse of myself, passing by a mirror or a shop window, my immediate first thought is 'gee, I wish I looked like that.' Then I realize it's me, and the self-deprecation kicks in."

If a therapist continues trying to solve psychological problems only through his patient's conscious understanding, therapist and patient will be frustrated and disappointed. Because the patient continually regresses to the behavior that he knows, intellectually, must change, both therapist and patient believe themselves to have botched the therapy.

Patients perceive the information the therapist offers not only at the conscious level but also at the pattern level, and what the patient hears may stimulate patterns of which neither the therapist nor his patient may be aware. The incoming information often is misunderstood or blocked altogether when it encounters a long-established pattern, so that the patient's reaction to new information or new strategies can be unanticipated and inappropriate—a monkey wrench that throws the therapy off-course.

Say, for example, the patient is a woman whose husband bullies her. Her therapist thinks that after just three appointments he has succeeded in helping her understand that it is her husband, not she, who is responsible for the bullying. Now the therapist broaches the subject of change, telling his patient that although she can't change her husband's behavior, she can change her responses to it. "You're responsible for changing yourself," he tells her.

She nods; she apparently agrees.

The therapist thinks that he is saying rational things to a rational patient, and that his words have struck home. What they actually have struck, though, is the "Play" button on an old tape   recorder buried deep in the patient's psyche. That word "responsible." The patient is hearing not her therapist but her father: "Of course you can't have a puppy. You're too irresponsible to take care of it. You lost your tennis racquet and now you want me to buy you a new one, huh? You'd lose your own head if it wasn't attached to your shoulders. No, I won't pay for you to go to college— you're too irresponsible to get yourself out of bed in time for class. The only thing for a girl like you is to find some man who'll take care of you."

Irresponsible. Irresponsible. You can't. You can't. You can't take responsibility. You can't take care of yourself: that is what the woman is hearing. What's more, she is perceiving her therapist's well-intentioned words as judgmental and critical, just like her father's. She becomes uncomfortable and balky, like a child. The therapist is disappointed. Unless he is sensitive enough to probe the meaning of her distress and break the impasse, or unless the patient actually speaks up and tells him what is upsetting her (unlikely, as she is already beaten down, and the tapes are so loud they muffle her instinct for self-preservation), this therapy session is going nowhere.

REPLACING PATTERNS

A GOLFER WHO tends to hook or slice may seek professional help to correct his defective swing. He knows something is wrong—witness all those golf balls landing in the rough, sometimes never to be seen again—but he can't figure out what. Cursing hasn't helped any, nor has hurling his club to the ground. His blood pressure is rising, and so is his handicap. Time to call in a pro. But before he arrives on the scene, let's take a careful look at this struggle between large frustrated human and tiny dimpled ball.

It's a brisk, beautiful spring morning when Bob, our golfer, arrives at the first tee with the other members of his foursome. He's feeling optimistic as he inhales the familiar smell of grass and gazes toward the carefully mown fairway. He chooses a driver, positions an orange tee, gently places a ball upon it. He addresses the ball, adjusts his stance, waggles the club a bit. It's the moment of truth. Bob's backswing is strong and graceful, the clubhead hovering a

374

millisecond before it begins its swift and mighty descending arc, and…and…*whoosh! whomp!* The ball erupts skyward, but Bob knows better than to look at it, for first he must complete the swing, his disciplined head still firmly bent toward where the ball was, his body twisted as though by a harsh wind, left foot planted like a great oak, right heel lifted balletically. Perfect swing, he tells himself. Perfect follow-through. *Masterful.* He looks up. There is his ball, black dot against cobalt sky, lifting, lifting, it is *Friendship 7,* it is *Apollo,* it is *Atlantis.*

No it isn't.

Short of its zenith, it yaws, pitches, yields to gravity, plunges deep into the rough.

"Tough luck, buddy," somebody says. Bob emits his first oath of the morning, scrabbles in the grass for the little orange tee, trudges away in search of his ball.

The work of pattern therapy is to replace aberrant patterns with new stimulus-response patterns that restore homeostasis. For pattern therapy to be effective, it must redirect the stimulus to some reaction other than the grooved pattern.

There are several ways to begin the process. The therapist may try, for example, to block the incoming stimulus from contact with his patient's response area. He may try to induce his patient to recognize the old stimulus as something new, so that the pattern filter allows it to penetrate and generate new neural patterning.

The more often this occurs, the less likely his patient is to identify the stimulus in its old form and funnel it to the old pattern. Or the therapist may try to block the patient's area of response, thereby modifying the actions that result when the patient recognizes the stimulus.

To clarify, let's return to Bob, who has sought treatment. He consults Gino, the golf pro, and tries to explain the problems with his swing. Gino listens, nods sagely, and sells Bob a book and its accompanying video. *Five Foolproof Steps to a Better Swing*, by Gino Abruzzi. $35.95. "Now, I want you to stay away from the course for a week," he tells Bob. "Read the book and watch the video as many times as you can. Then see me in my office. We'll get you straightened out."

Bob follows Gino's advice to stay off the golf course. Thus the operative stimuli, golf shoes, golf bag, clubs, ball, fresh air, smell of grass, are blocked. He reads the book twice and watches the video every night after dinner, thereby acquiring a new approach: he begins to view the difficulty with his swing as a challenge to his mind rather than to his body. In other words, he sees the old stimuli as something new. Bob is thrilled. The week passes. He keeps his appointment with Gino.

The office walls are gray cinderblock. The floor is bare concrete except for a mat made of stiff plastic grass. "Let's see your swing," Gino says. "Grab one of those drivers over against the wall." Bob takes one and steps onto the mat with it. The plastic grass gives off an acrid odor. There's no ball, no tee. The unfamiliar driver feels strange in his grip. Self-consciously, Bob addresses the spot where a ball should be, swings.

"Again," Gino orders. "Keep at it."

He slowly circles Bob as he swings, over and over, at an invisible ball.

"Okay, Bob, I see the problem. At the top of your backswing your right shoulder is crowding your ear, so when you bring the club down you have to loosen up or you'll wrap the club around your knees. But you overcompensate, you get a little too loose, and you lose some control of the club. We're talking tiny increments of movement, understand, but it all counts. Here, let me show you what I mean."

He takes the club and performs a slow-motion imitation of Bob's swing. "Now watch how I do it."

He executes another slo-mo swing, silken and perfect.

"See the difference?"

Yeah, Bob thinks gloomily.

It's the difference between the Hunchback of Notre Dame and Tiger Woods.

After an hour of implementing adjustments to his swing, Bob heads for home and a week of nightly practice in his basement. ("Stay off the course. Just keep practicing what I've told you.")

Very shortly, the new, improved swing becomes his norm. He can scarcely even remember how to swing the old way. In a week or two, equipped with new knowledge, he'll graduate to the practice tee.

The fresh air and the smell of real grass will evoke only pleasure, not bad memories. The golf ball will soar as it should, straight and true, and if it doesn't, Gino is by his side to interject advice and encouragement before Bob has time to get flustered.

Besides, he now not only *knows* how to correct his swing, but at a deep, satisfying level he *feels* that he can. Thanks to psychologist-without-portfolio Gino, after one or two more lessons Bob will be on his own.

THE USE OF DRUGS
IN PATTERN EXPERIMENT

CORRECTING A GOLF swing is a whole lot simpler, of course, than correcting a lifetime's worth of deleterious psychological patterning. These days most therapists employ mixed methodologies, often a combination of talk therapy and psychopharmacology. The patient-therapist dynamic is the foundation of talk therapy in combination with medications.

There is an enormous variety of drugs to augment the talk therapy process: antipsychotics, antidepressants, stimulants, and anti-anxiety medications which can help a patient perceive a stimulus differently and modify his pattern of response. The hope is that repetitive modification of the stimulus-response that occurs while a patient is taking the drug will result in permanent change to the malfunctioning pattern within a fairly brief time, after which the patient can eventually stop using the drug.

Medication is especially helpful for people with chemical imbalances that cause mental illness. At the same time that these drugs are addressing imbalance issues, they can serve to block old patterns and allow development of new ones. Drugs also are useful in enabling seriously neurotic people to function reasonably well.

Unfortunately, no real "cure" is occurring because drugs don't alter patterns; their purpose is to make it easier for patients to do the work of change. What really has to happen is that the patient gains insight into his perceptual patterns and the responses he has built so that he can set about creating blocking patterns.

Illicit drugs as well as alcohol have a long and close association with pattern-breaking. Young people, particularly young men, may frequently use drugs to build up the energy needed to break out of childhood dependency and establish independence. The new patterns are thus drug-generated. Even if the user withdraws from the drug, the drug-induced patterns have been established, and they remain in place along with the patterns that impelled the person to turn to drugs initially. Then, under the right—or wrong—set of circumstances, the familiar drug stimulus may elicit more drug use, and the individual may be well along the road to addiction.

Legions of creative people, writers, musicians, artists, have used drugs to open up perception pathways and alter response patterns. Like many other writers of his time, Coleridge, for example, smoked opium. One outcome was the extraordinary "Kubla Khan"—which, according to Coleridge, came to him during an opium dream. Aldous Huxley experimented with mescaline.

In 1954 Huxley published *The Doors of Perception*, detailing the insights he experienced while using the drug, which ranged from the "purely aesthetic" to "sacramental vision." Much of American jazz has partial genesis in

380

musicians' use of cocaine and heroin. The long roster of artists and writers who are known to have favored absinthe as both impetus for, and respite from, their creative labors includes, among many others, Van Gogh, Toulouse-Lautrec, Modigliani, Verlaine, Baudelaire, Wilde, and Hemingway.

And who can forget the "psychedelic" music and art of the Sixties? Given such an abundance of creative talent and the interesting variety of substances that seem to have fueled it, you're almost inclined to try some yourself. On the other hand, Salvador Dali, whose surreal paintings practically scream of a world accessible only through dramatically altered consciousness, insisted that he abstained from drug use.

"I don't do drugs," Dali said. "I am drugs."

Drugs may alleviate some kinds of feelings, but they don't obliterate patterns. If science ever discovers a drug that eliminates the destructive grooves themselves, there is the risk that it also will destroy the necessary, normal grooves that people need to continue functioning. For therapy to be effective, the patient must be re-grooved; the old tapes need to be remastered.

PSYCHOTHERAPY IS HARD, intense work. Emotions, such as love, hate, fear, shame, are the glue that holds pattern modifiers in place. Military boot camp, for example, uses fear, fatigue, competition, and peer-group approval to modify the patterns of young people entering the service. Nine weeks of boot camp can transform a spoiled, sulky, overweight underachiever—a kid no one expected ever to amount to anything—into a hard-muscled, skilled young soldier with a purpose: to accomplish the mission. Love is another strong agent of change. Consider for instance the young pregnant woman who, for love of the baby she is carrying, has given up cigarettes and wine and even her customary morning mug of coffee.

The strongest pattern changer is the sincere desire to change. First, though, the individual must be persuaded not just that change is necessary, but that it will enhance his life. You and I may clearly see that a person needs to change his ways, but it probably isn't so obvious to him, and what's more, the prospect is frightening. If, however, he can be brought to understand that change will benefit him, his own self-interest can eliminate or block old patterns in a way that makes it difficult for them to reassert themselves. Here is where the therapist-patient relationship is most constructive.

In pattern therapy, the emotional energy that develops between therapist and patient can serve to change patterns. Sometimes it is a matter of the patient's realization that he and his therapist, bonded in a relationship of mutual respect, are working together to solve a problem. Also, sometimes change is effected through epiphanies or these moments of often highly emotional insight that clarify what was formerly muddled and confusing. Properly managed by a caring therapist, these epiphanies can block or modify a patient's typical response to a familiar stimulus. The insight releases energy that aborts the pattern by short-circuiting the stimulus and re-routing it to the conscious mind, where the stimulus must call out a new, as yet un-grooved response.

With each repetition, a new pattern is formed and increasingly strengthened until genuine change has been established.

Epiphanies are exciting. Just ask the Three Wise Men who for twelve days and nights of hard traveling followed that Star. Or, ask Sarah, the woman I mentioned earlier who was so alarmed by the prospect of change. "I had one of those epiphanies," she told me. "When my therapist explained how incest shatters a child's sense of self, how the destruction can define her character and blight her attempts to form lasting, genuine adult relationships, it was new information for me.

"What Dr. Morris said made *perfect* sense to me—it was the piece of the puzzle that let me begin to see the whole picture. It turned my world upside down, or rather, right-side up, for the first time in my life. Driving home from the therapy session I was so excited, so *thrilled*, really, with my new insight, that I was barely conscious of my surroundings. I ran a stoplight—the kind policeman just gave me a warning—and a few blocks later I ran out of gas."

It is important to remember that epiphany doesn't necessarily transform into consistent progress. In fact, as energy flags, the patient is likely to regress. Both patient and therapist must work to maintain energy. Think of Bob the golfer. With his new, improved swing Bob has won the club championship, and the energy that drove him to change his swing pattern and that gave him his competitive edge is abating. While he is resting on his laurels, the old pattern insidiously resurfaces. But with a couple of remedial lessons from Gino, the problem is solved.

In psychotherapy, both therapist and patient must work to sustain energy lest the patient regress too far. Yet the therapist needs to be aware that there is some advantage to these "down" periods. It seems that they protect the patient from taking on more pattern change than he can manage at one time, which can lead to frustration and failure. A good therapist recognizes these shifts in energy, and knows when safely to apply pressure, and when to let the therapeutic process simmer gently for a while. Psychotherapy is an art, a science, and a very delicate balancing act.

CAGES AND SOCIETY

IN THE BEST possible world, pattern therapy might also include cage therapy, as people inhabit various cages—self, family, company, country—and in each cage they respond only to what they are patterned to perceive. By instituting the practice of family therapy, psychology has made a small step in the right direction. Society comprises a daunting multiplicity of cages. In dealing with societal illness, we need to develop techniques that improve perception among caged groups. That means arduous work, and only the intrepid need apply.

New patterns must not only be established, they also must be maintained, or they will atrophy.

Is the idea of universal pattern therapy no more than a pipe dream? Maybe, but it does seem like something worth trying. But how to implement the therapy on such an enormous scale? I would suggest starting with young children, who haven't developed programming into the beliefs and practices of their society. Young children have still-flexible minds and are most amenable to, and capable of, exploring new avenues of thought.

People love to say that children are our future, but few seem willing to equip them intellectually to create a *better* future. Rather, the civic education of young children often consists largely of indoctrination. And yet, children are naturals at critical thinking. Why not, early on, teach and encourage them to engage in it? Possessed of this essential intellectual skill, children are more likely to make good choices—first for themselves, later on for their families, their countries, and indeed, as succeeding generations of intellectually cooler heads prevail, for the world itself.

In an essay called "Group Minds," Doris Lessing offers a similar suggestion: *Imagine us saying to children, "[T]he human race has become aware of a great deal of information about its mechanisms: how it behaves, how it must behave under certain circumstances. If this is to be useful, you must contemplate these rules calmly, dispassionately, disinterestedly, without emotion. It is information that will set people free from blind loyalties, obedience to slogans, rhetoric, leaders, group emotions."*

And imagine us using pattern therapy to treat the ills of the world. It would seem a reasonable approach, as the individual patient so often mirrors the troubles of his society, where similar processes are at work. That is, old patterns function below the group's conscious level until new patterns are generated by force of energy, which frequently arrives in the form of pestilence, war, famine, and death.

Today, insofar as it addresses the wants, needs, motives, and psychological makeup of international leaders and their societies, diplomacy may be the nearest we have come to universal therapy. Sometimes diplomacy works, but more often the best we can do is try to stay one hoofbeat ahead of the Horsemen.

...AND ONE
LAST THING

I'M A REALIST. I don't foresee paradise on earth. But I believe we can do better. We can *always* strive to do better—as individuals, as a society, as a human collective, as a global force. Patterns bind us, they connect us, and they inform our futures as well as validate our pasts.

At the same time, patterns are everywhere—and they are everything.

Patterns of nature, of civilization, of history.

Throughout these pages, I've included the voices and thoughts of many classic poets. This says as much about patterning as anything I've said across the entire expanse of this book. Poetry is inherently unbound by patterns—there's an invitation to break conventions, confound rules, and challenge readers. The very purpose of poetry is to reveal patterns. Poets can either follow rigid, regimented rhyme schemes or veer off into delightfully unexpected directions with unrhymed lines and no fixed metrical patterns.

This isn't just the heart of poetry, but the heart of life itself.

When poets don't follow the rules, those rules become that much more apparent. We didn't notice the conventions until they weren't there any longer, much like not realizing what you had until it's gone. And that's not simply what makes poetry so vital and inspiring to me. It's a solemn reminder that patterns are always there, unseen in plain sight.

Patterns are the poetry of our lives, in both quiet and profound ways.